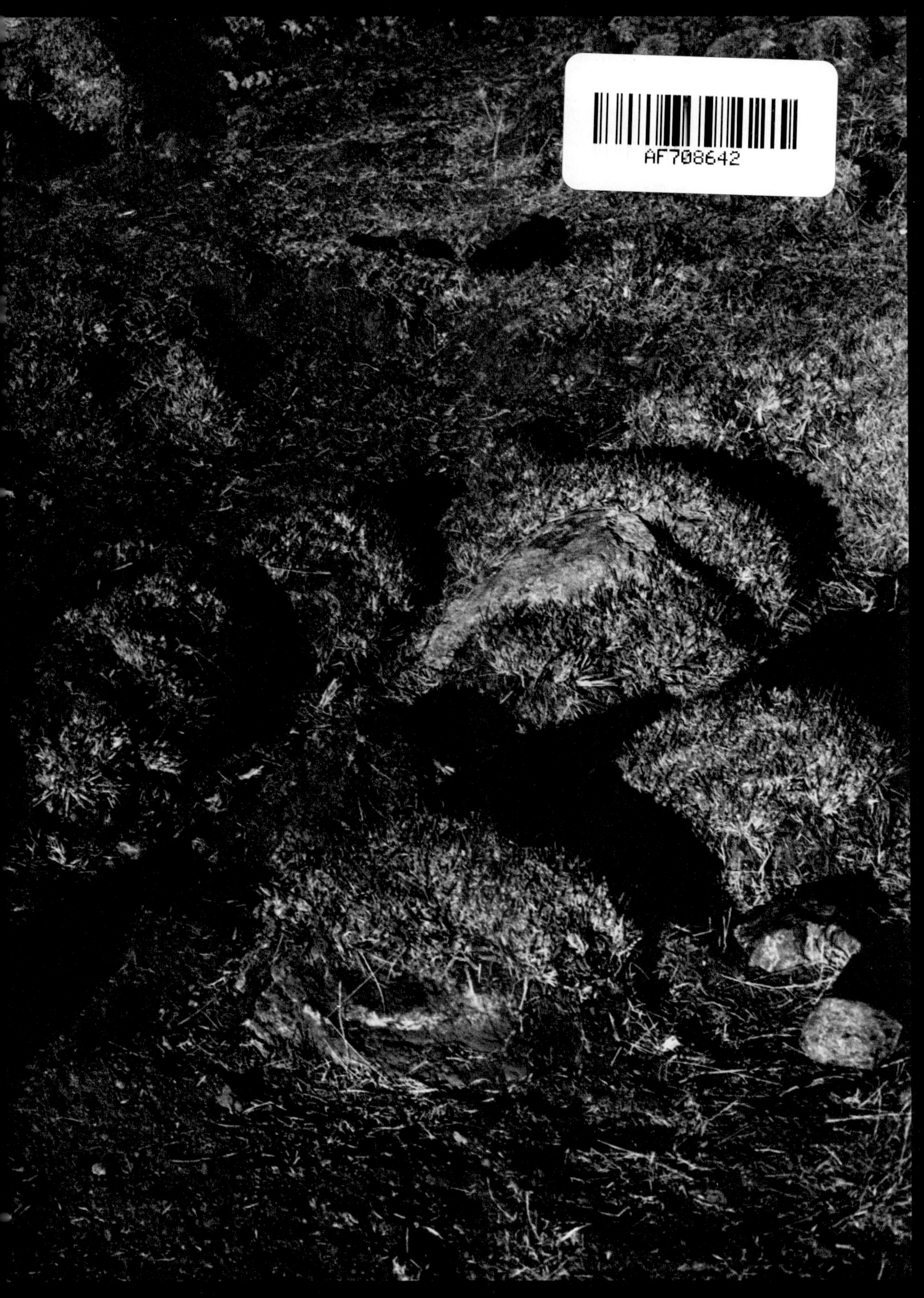

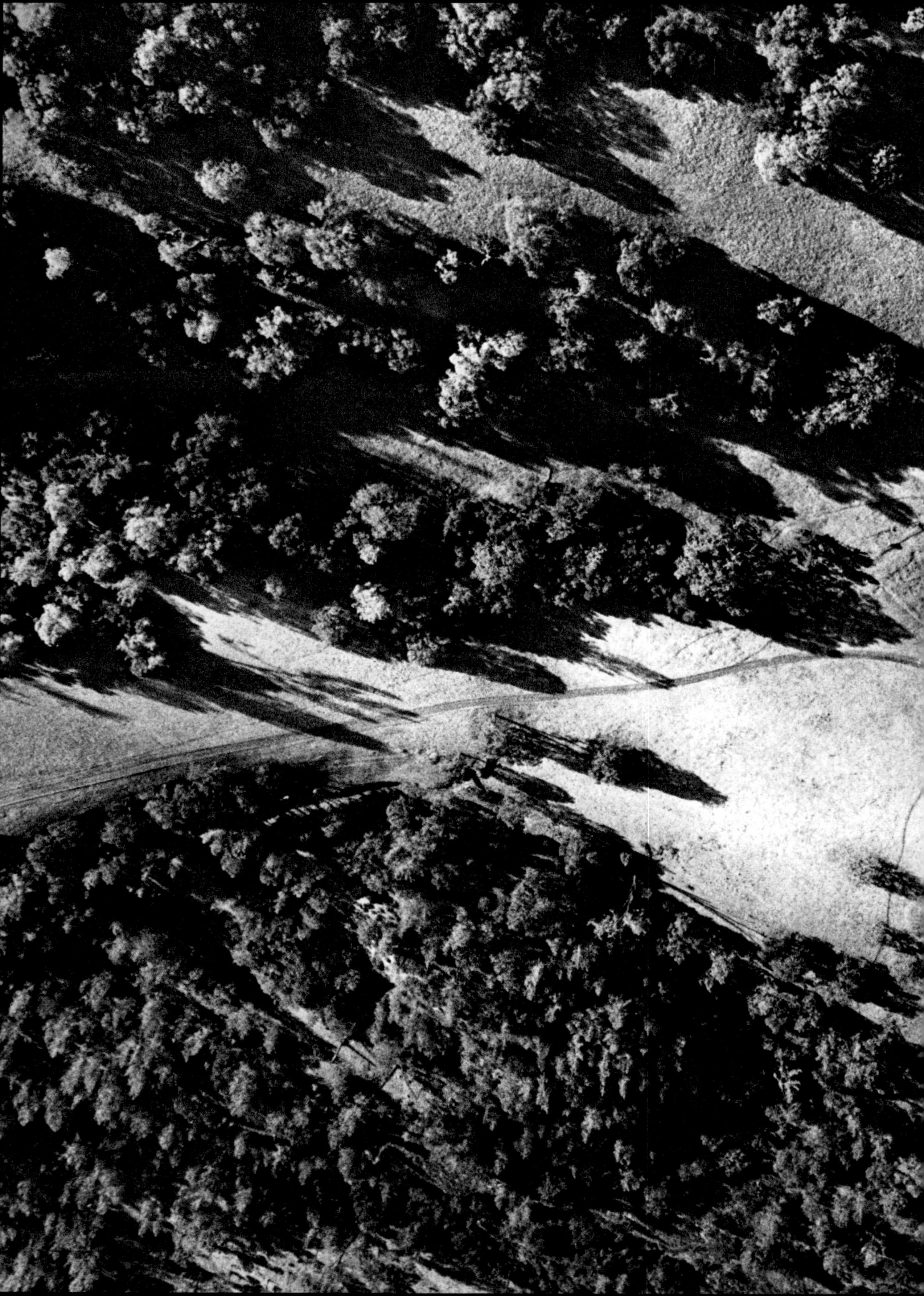

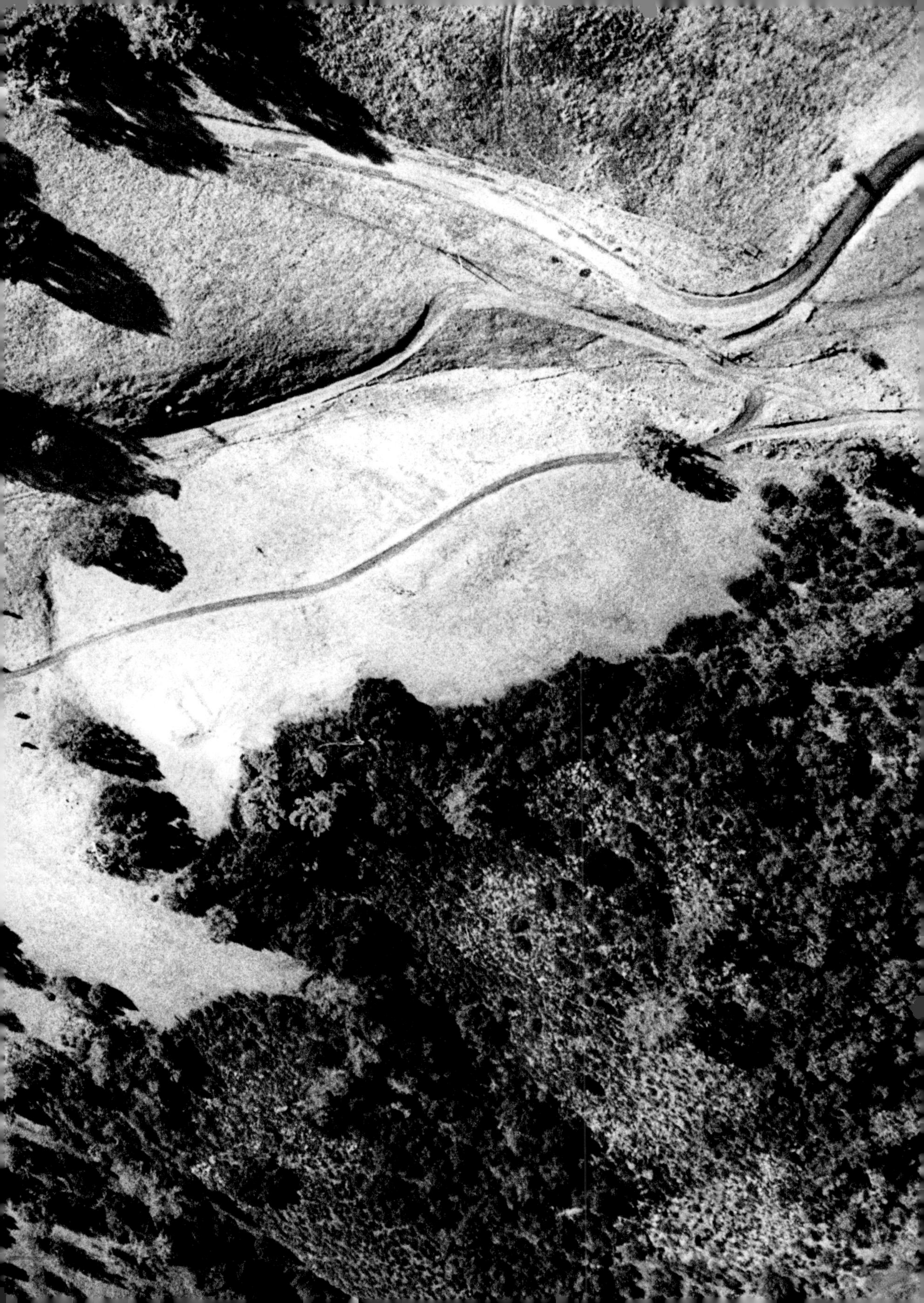

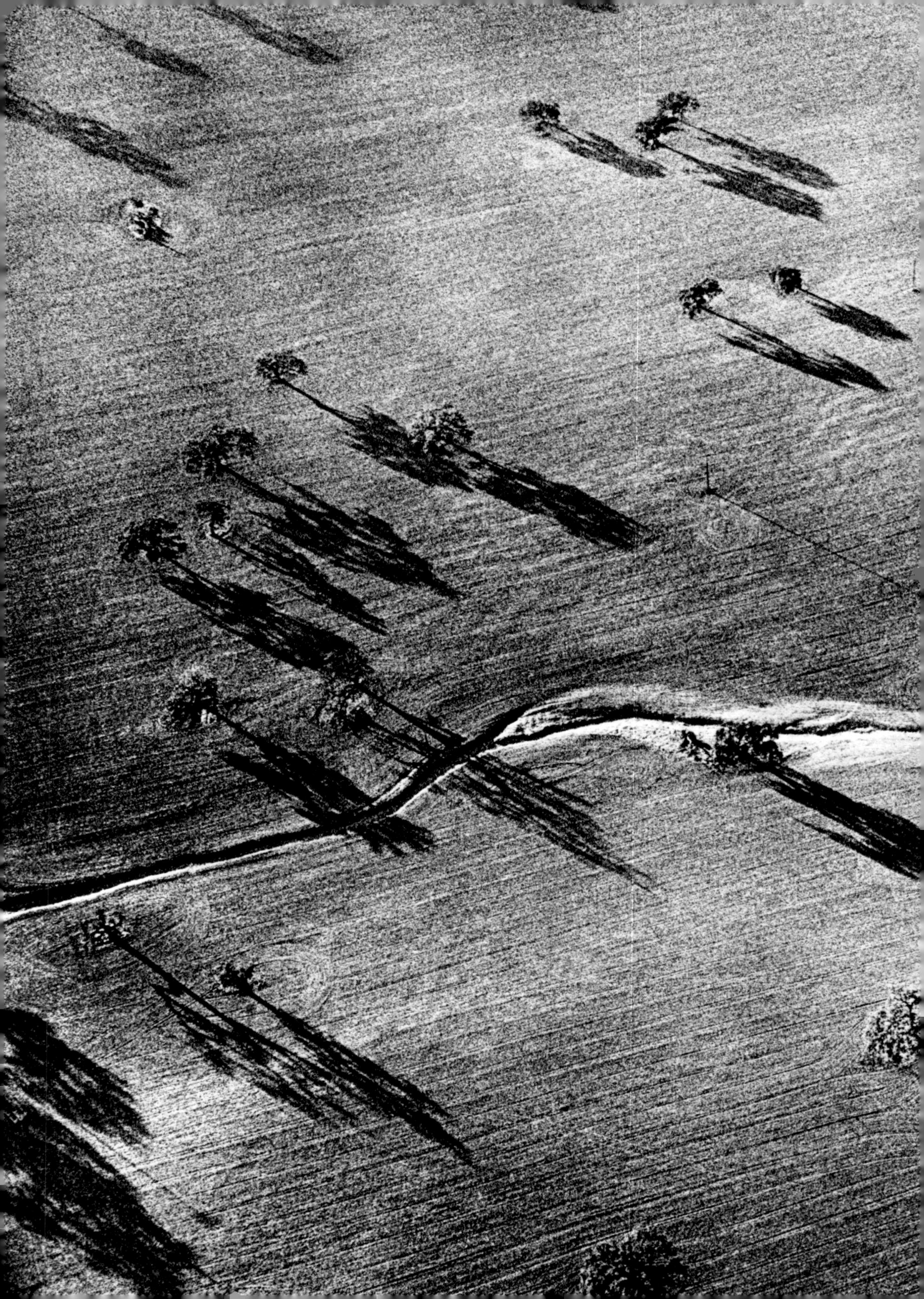

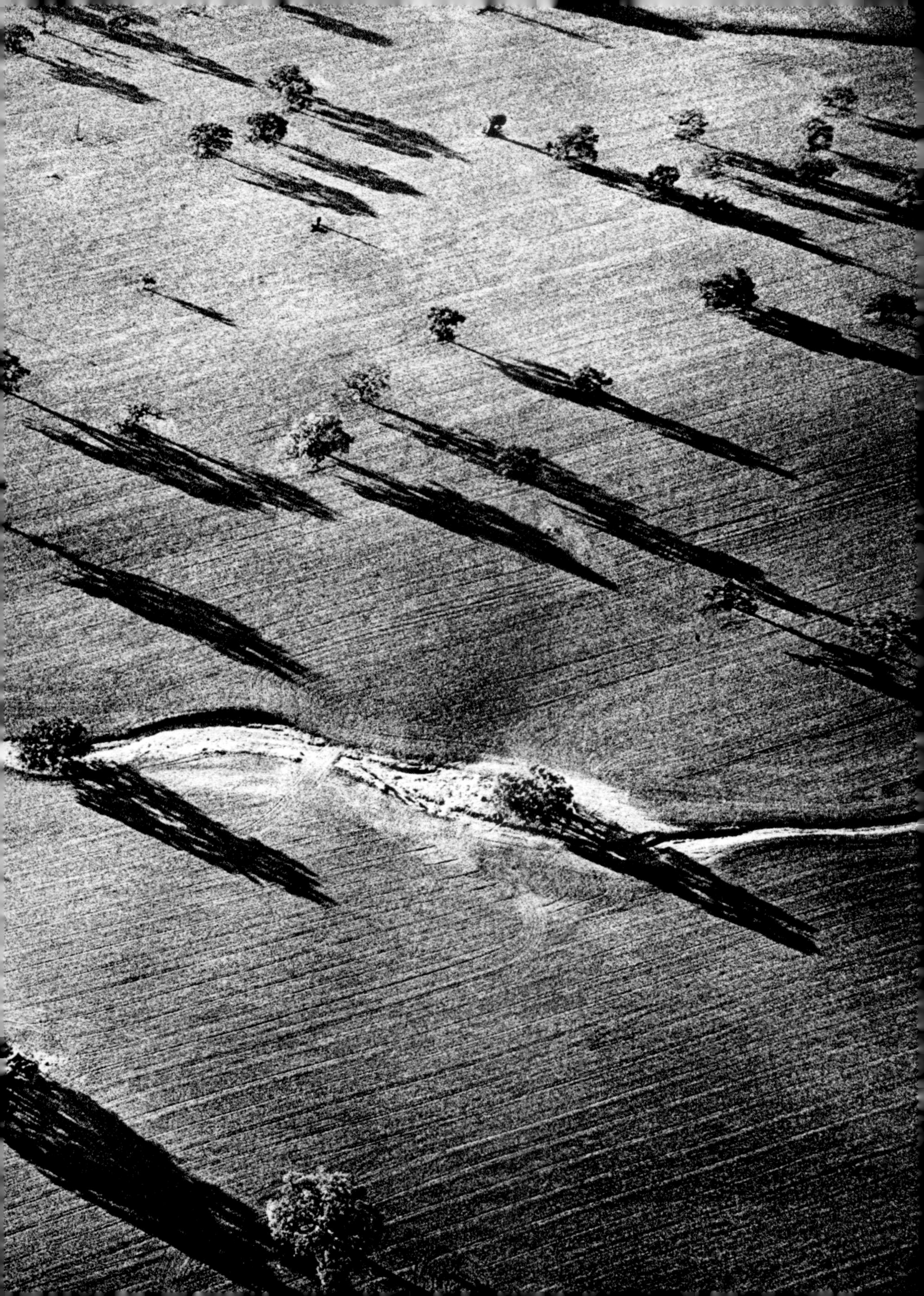

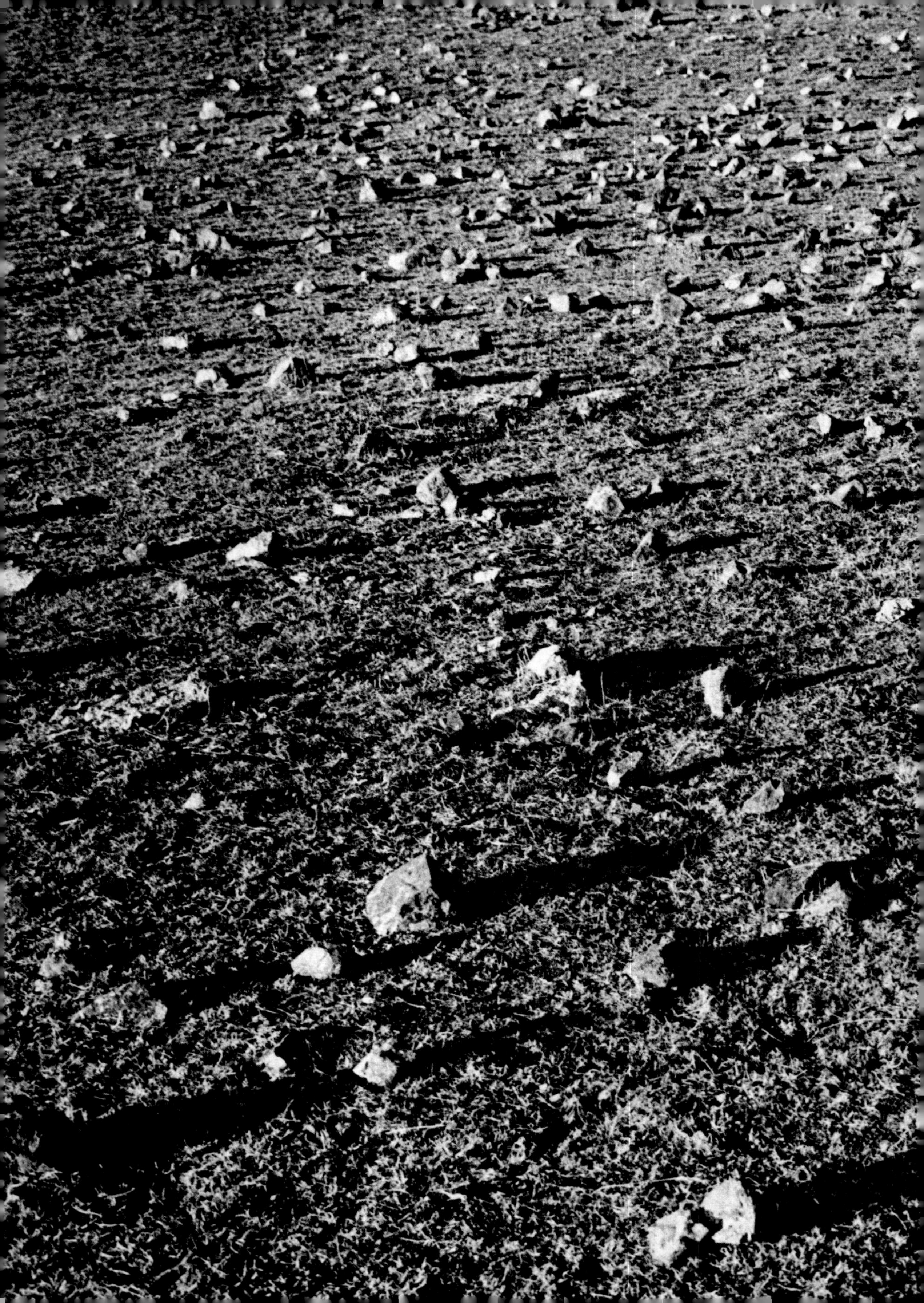

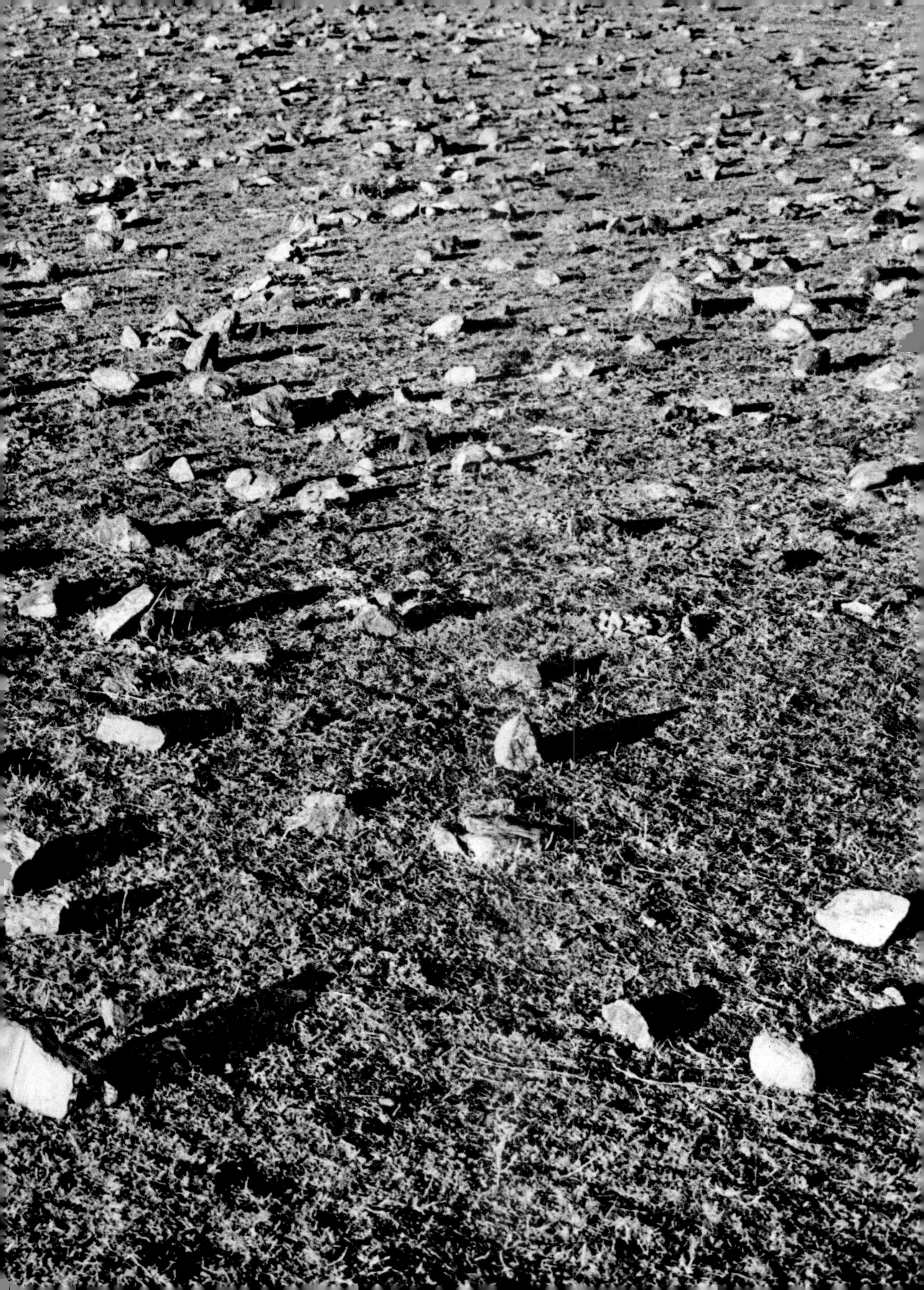

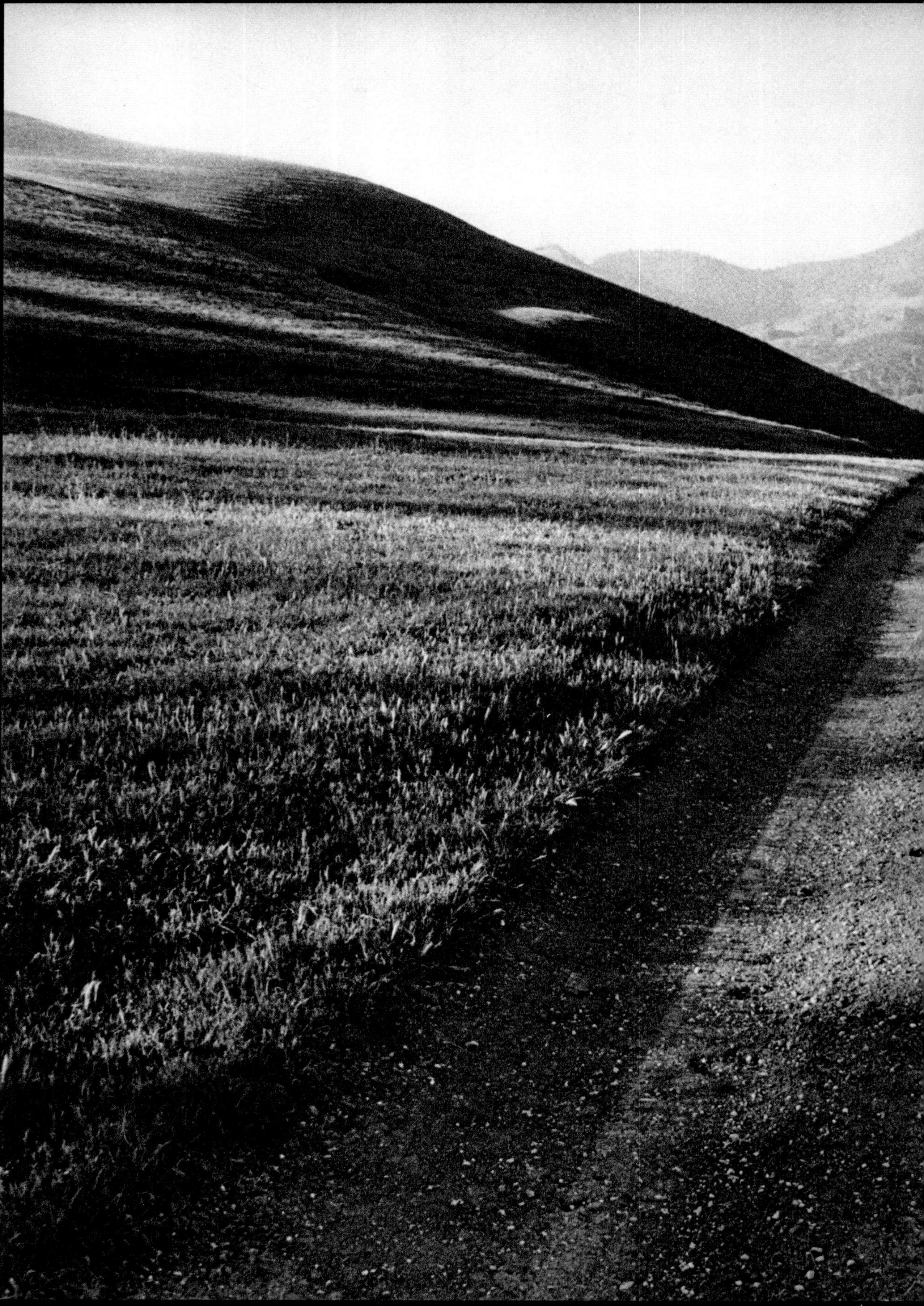

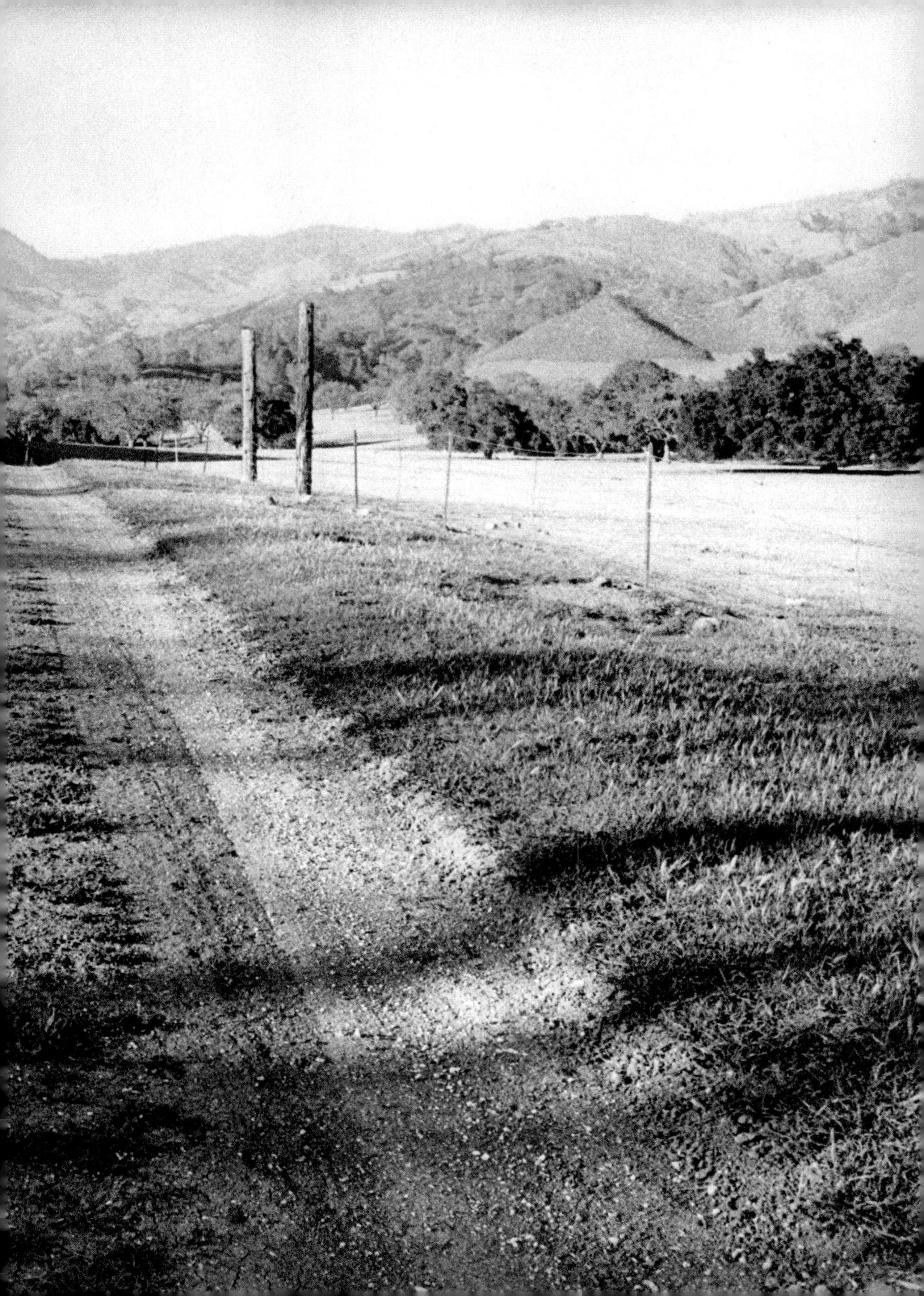

RANCH

MICHAEL LIGHT

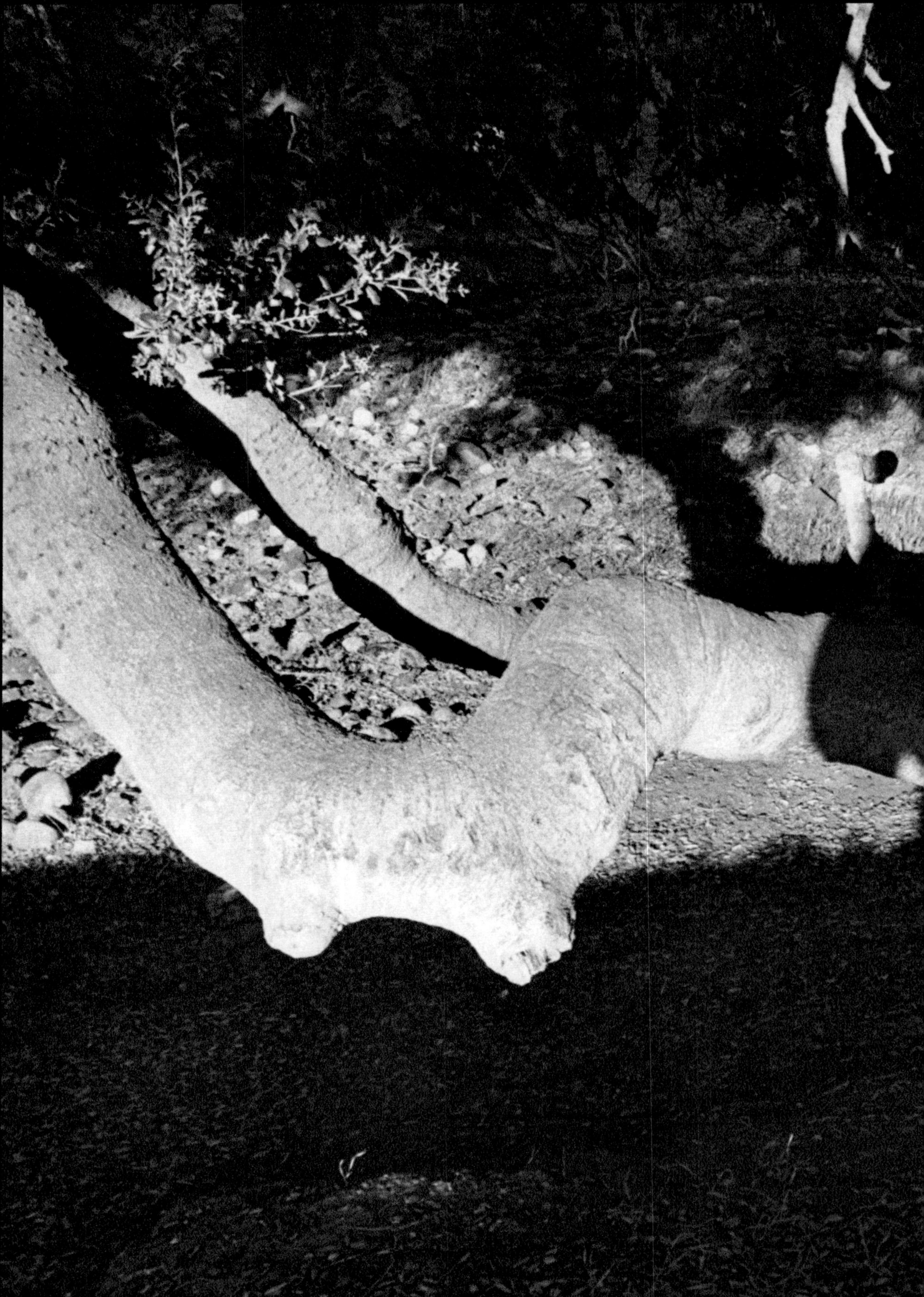

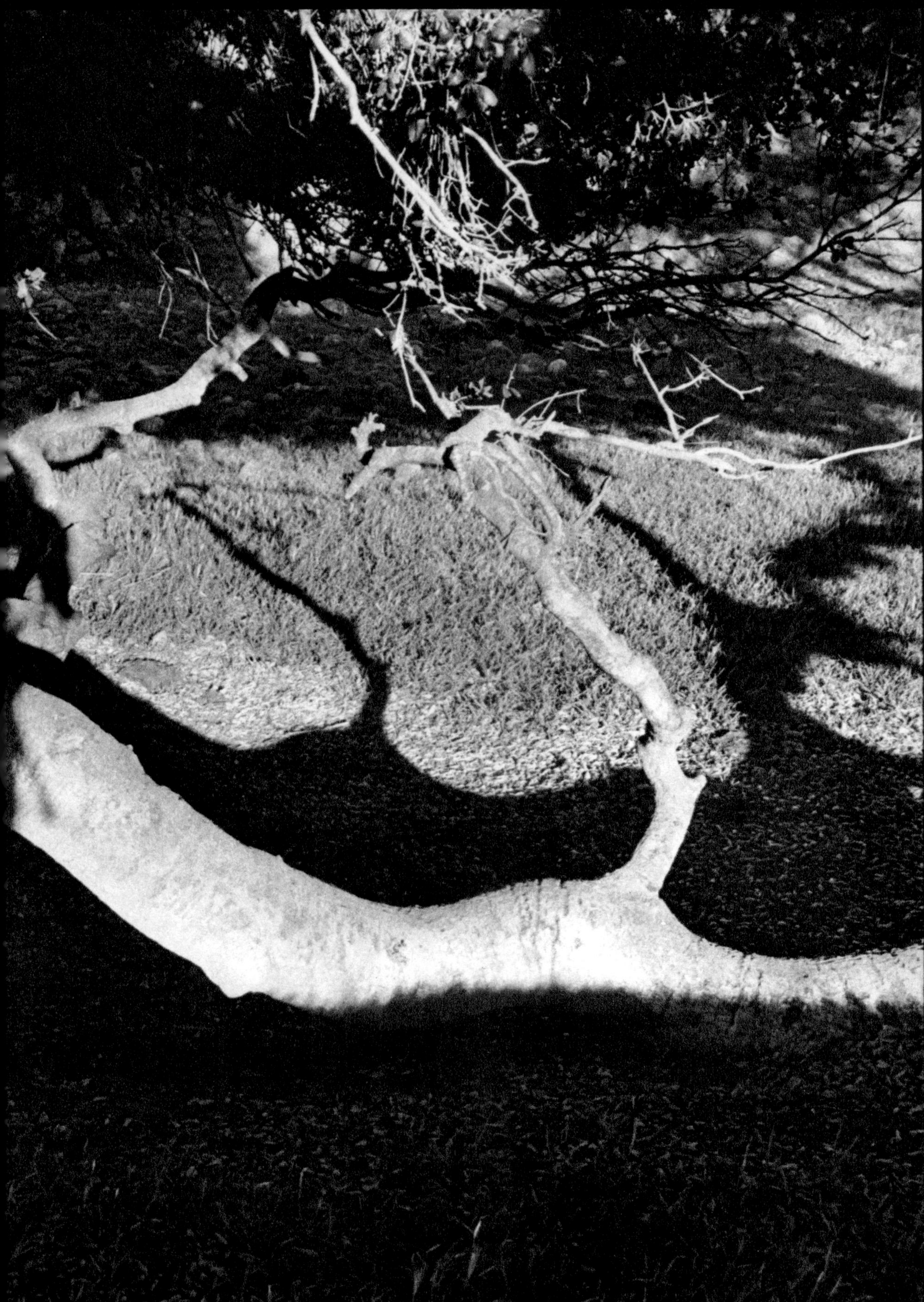

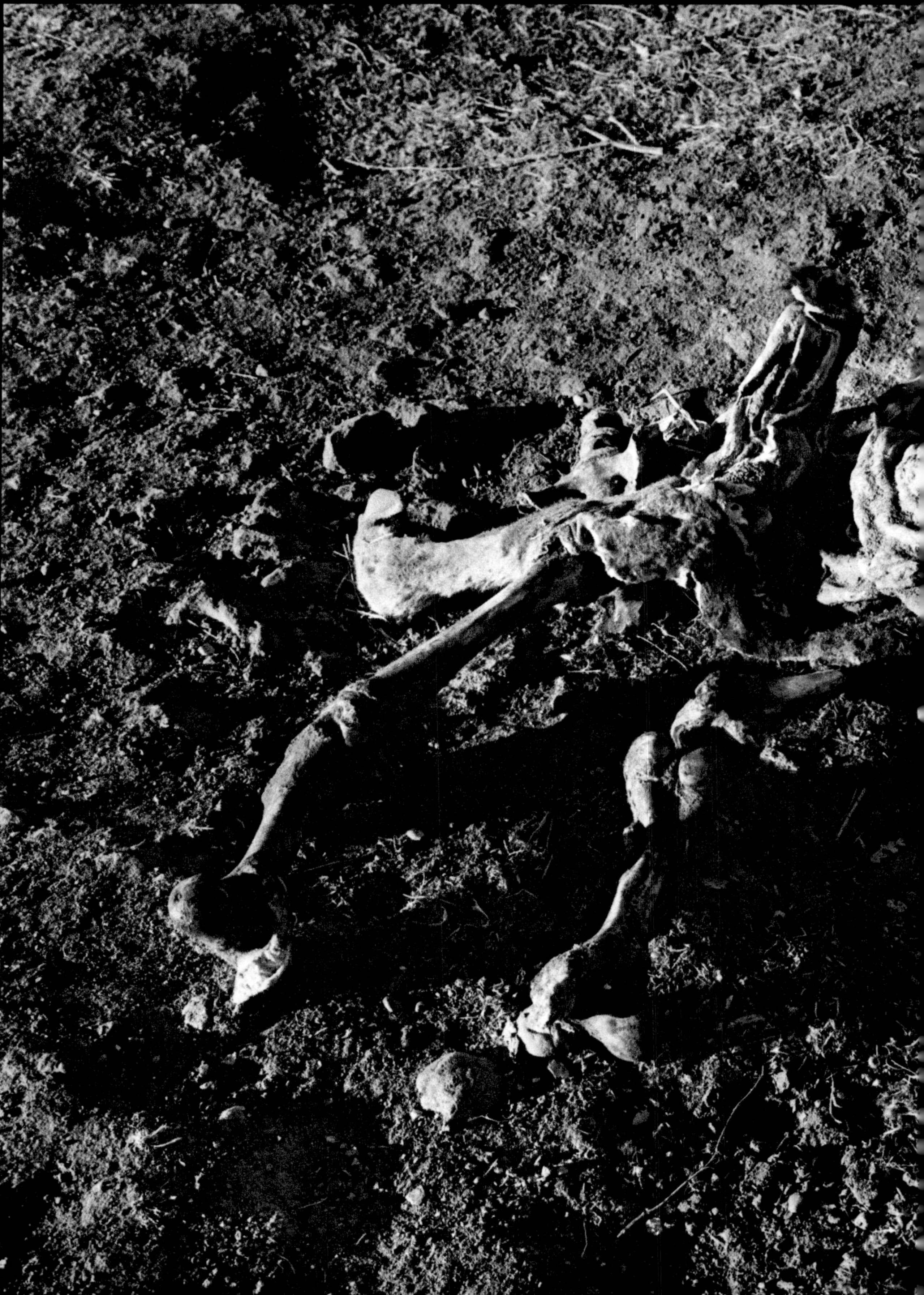

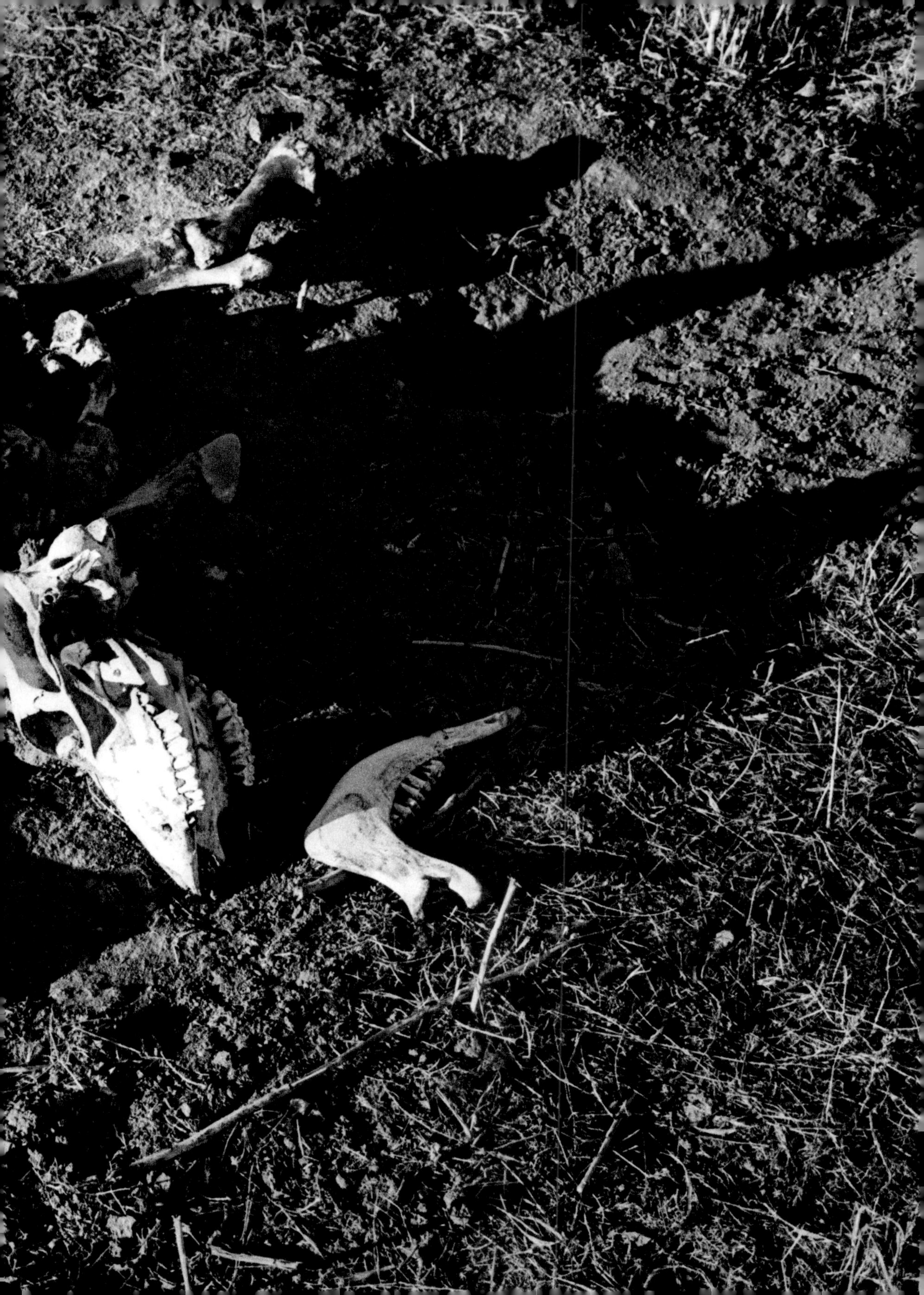

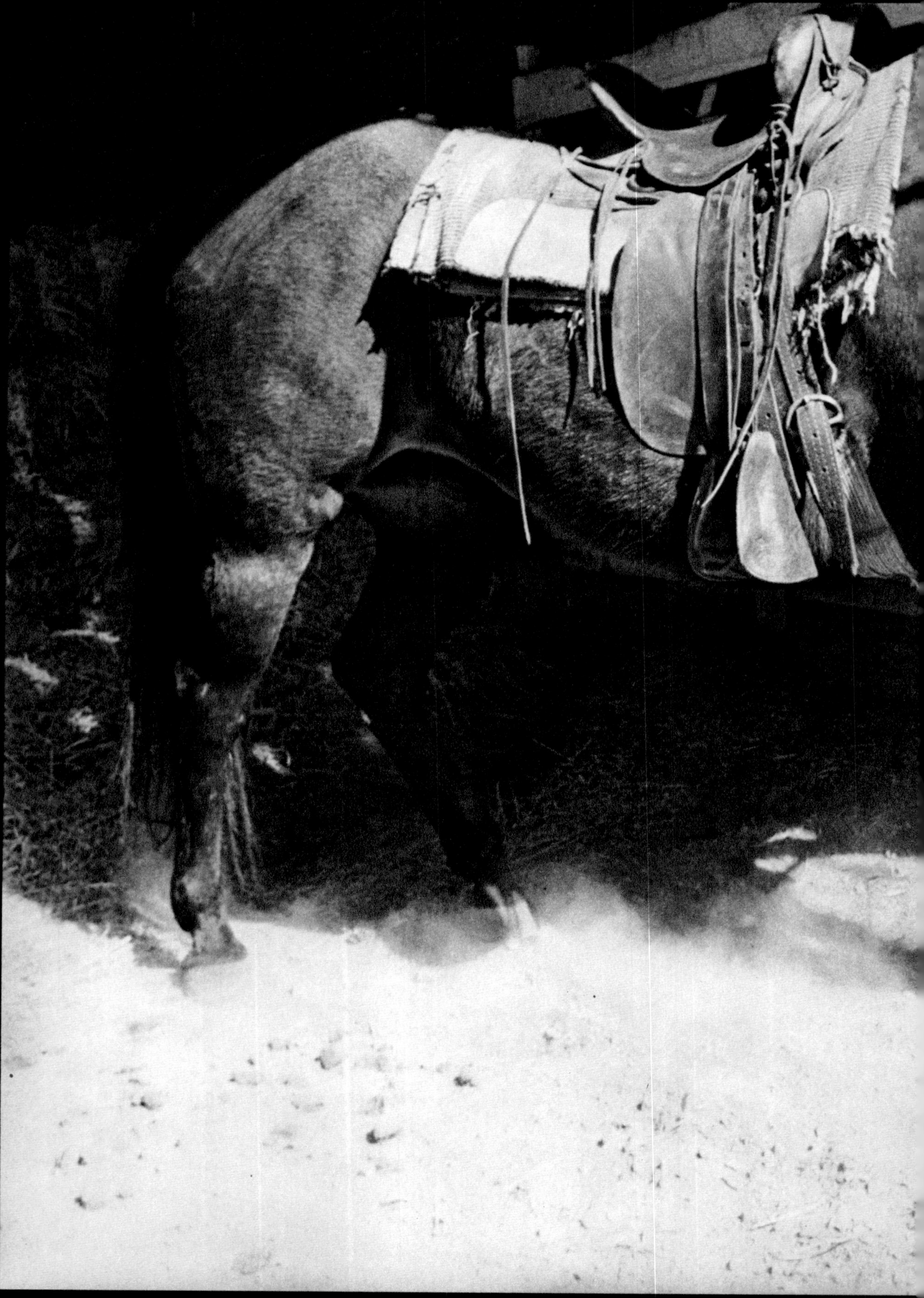

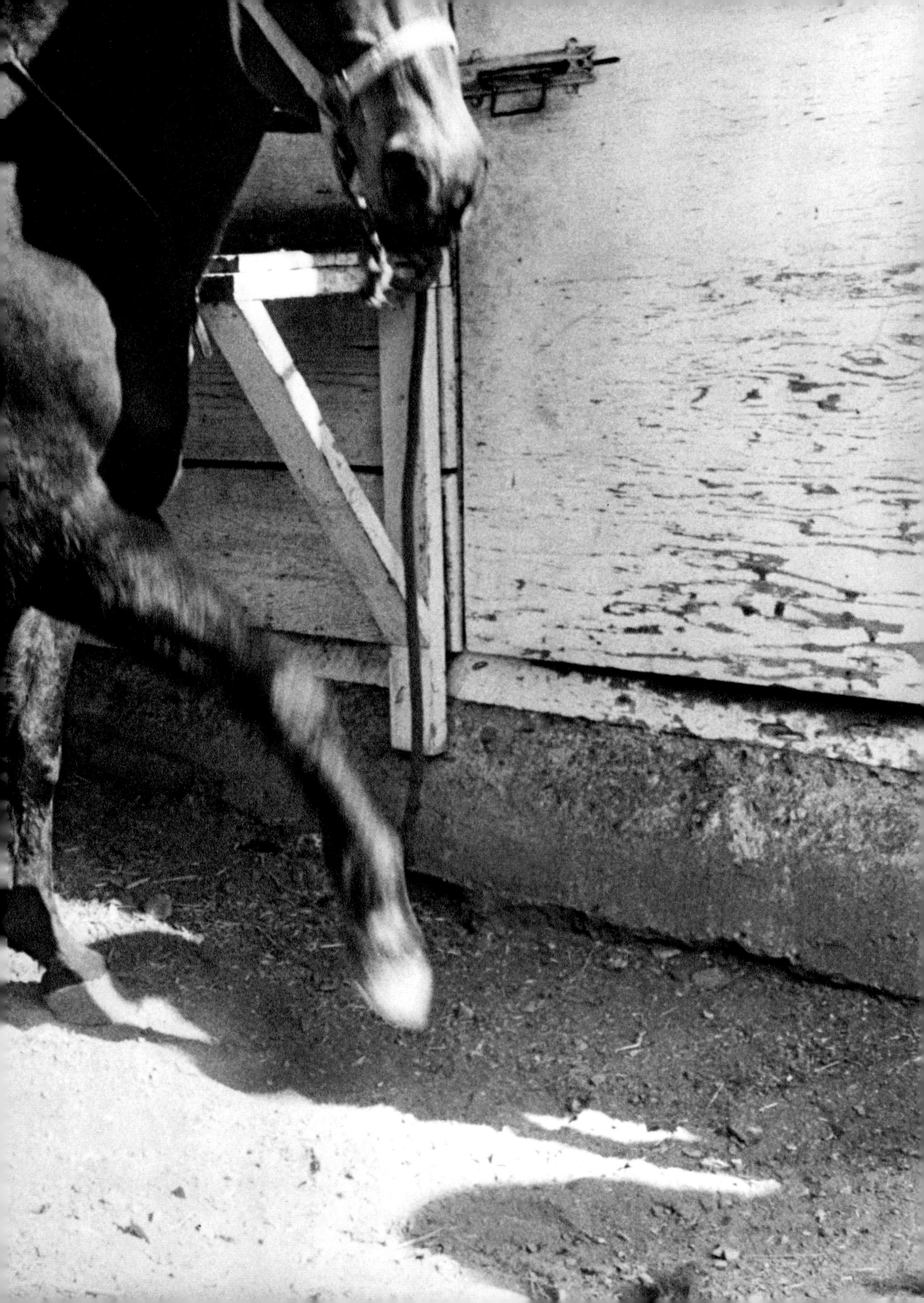

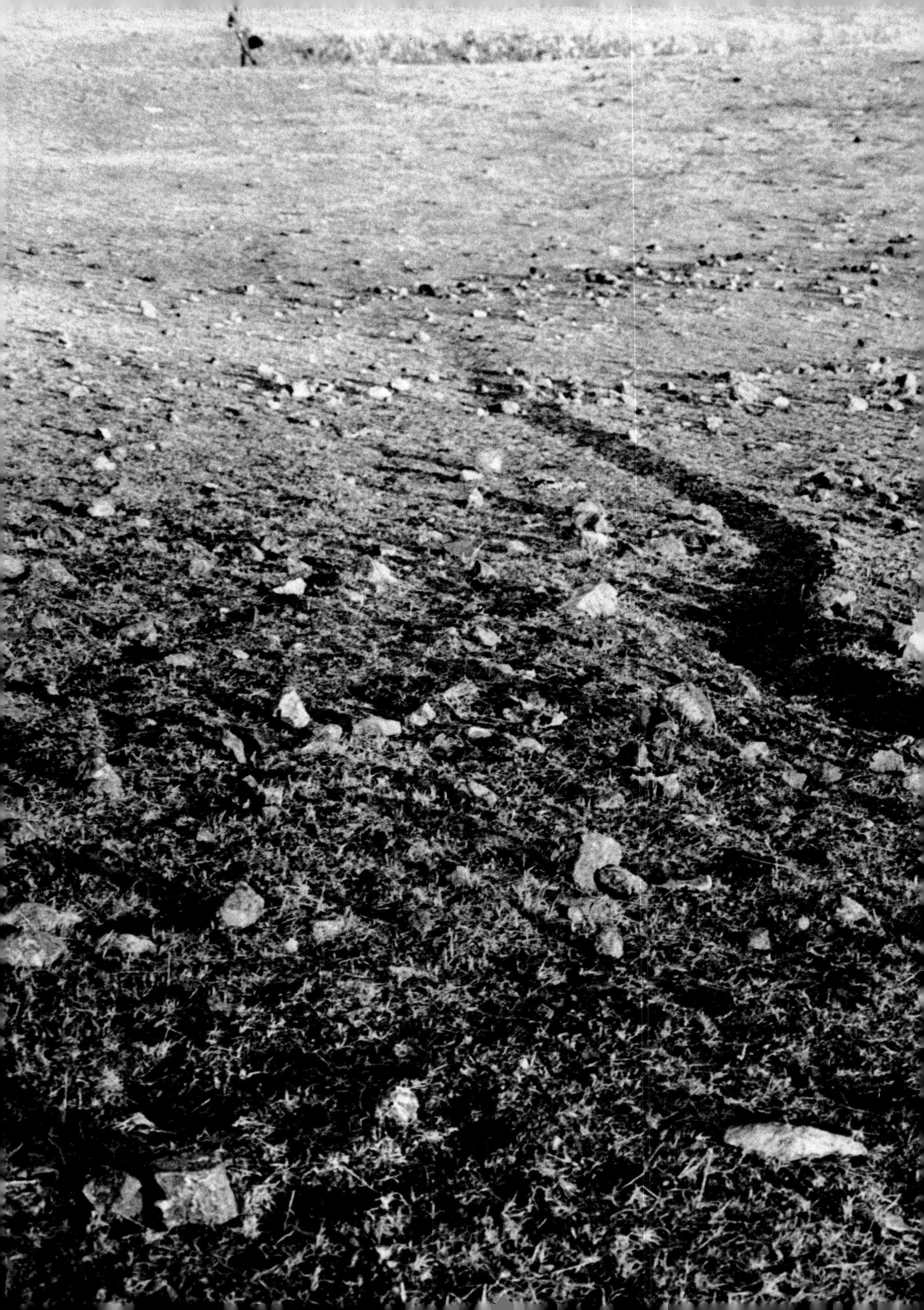

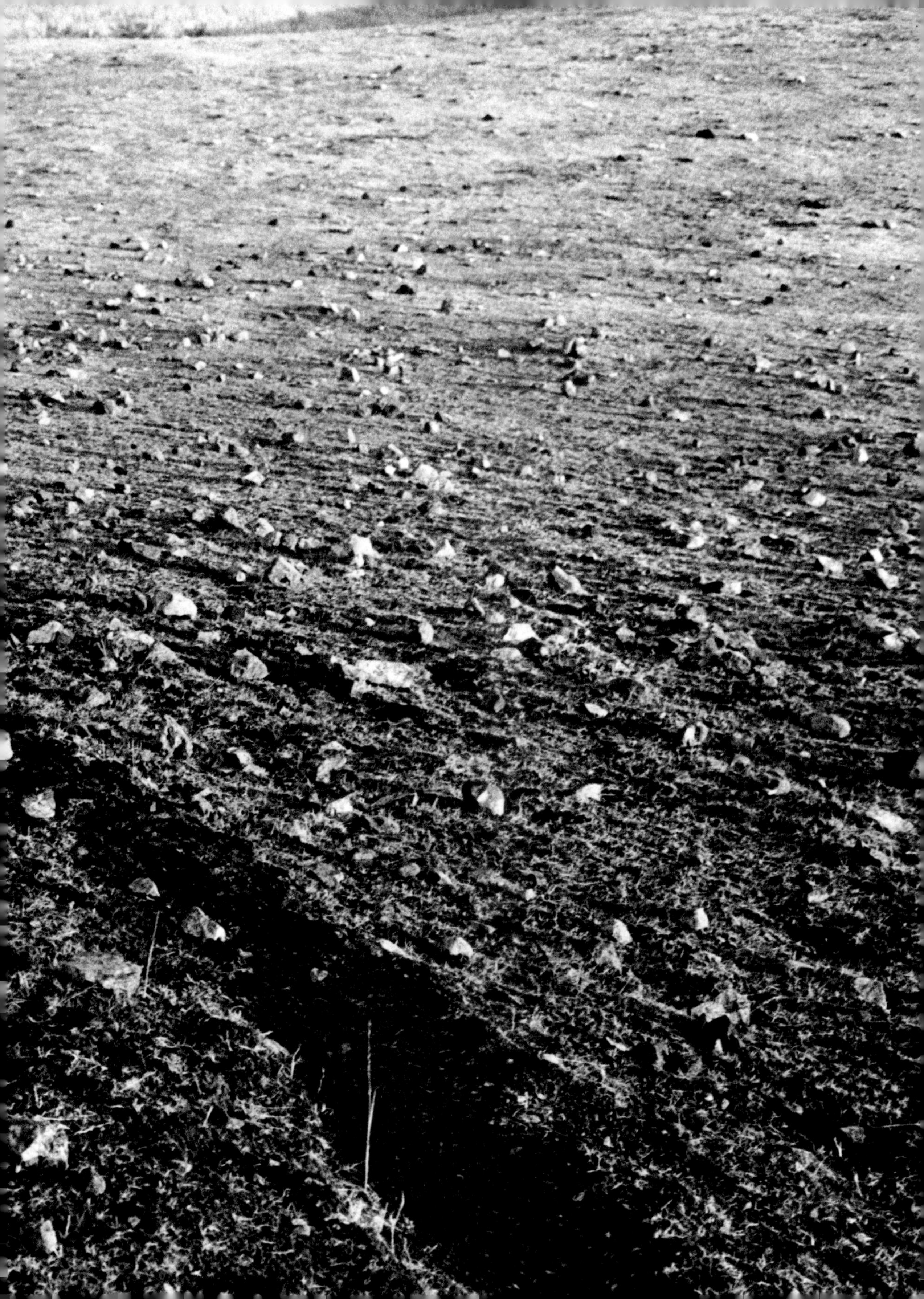

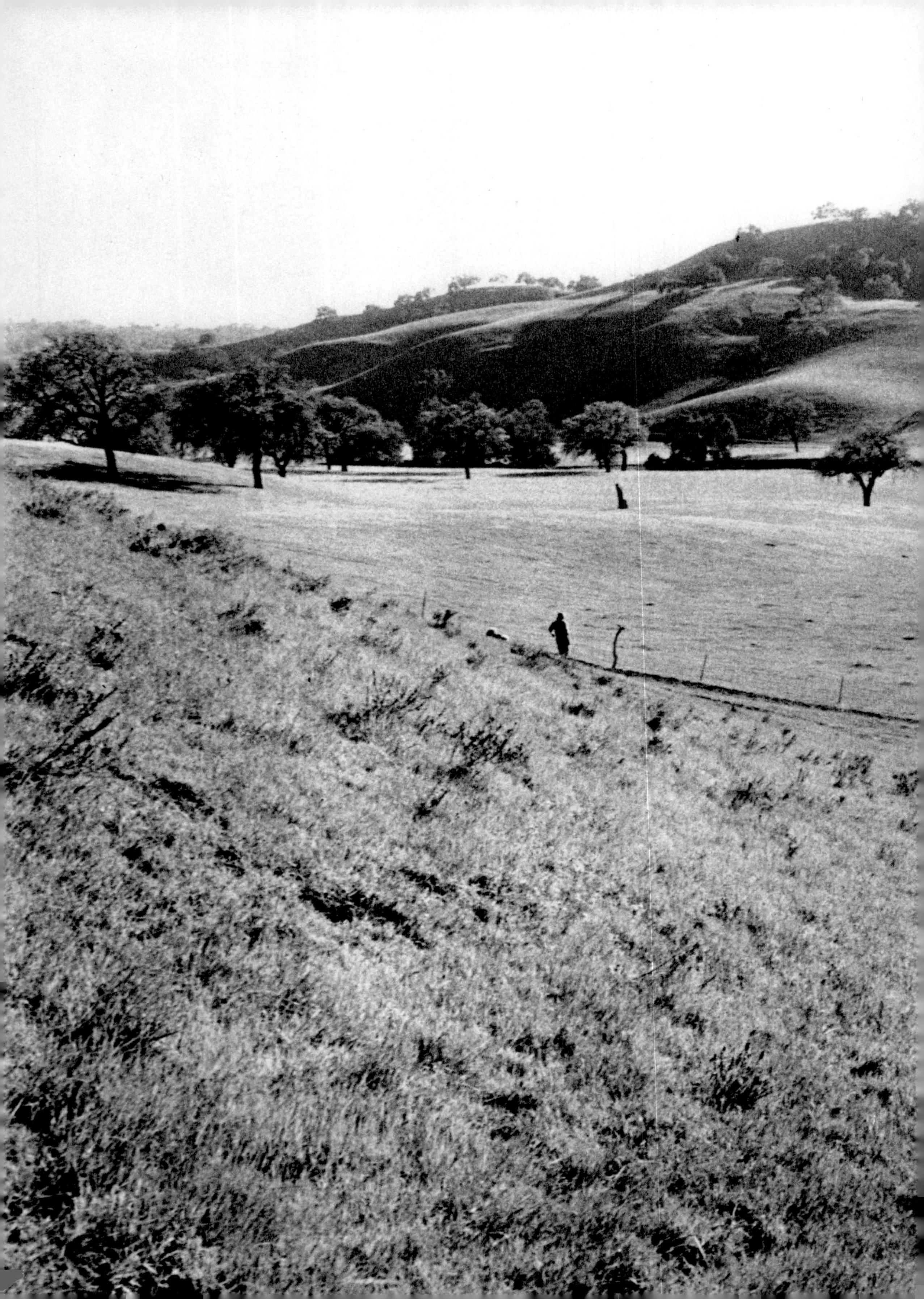

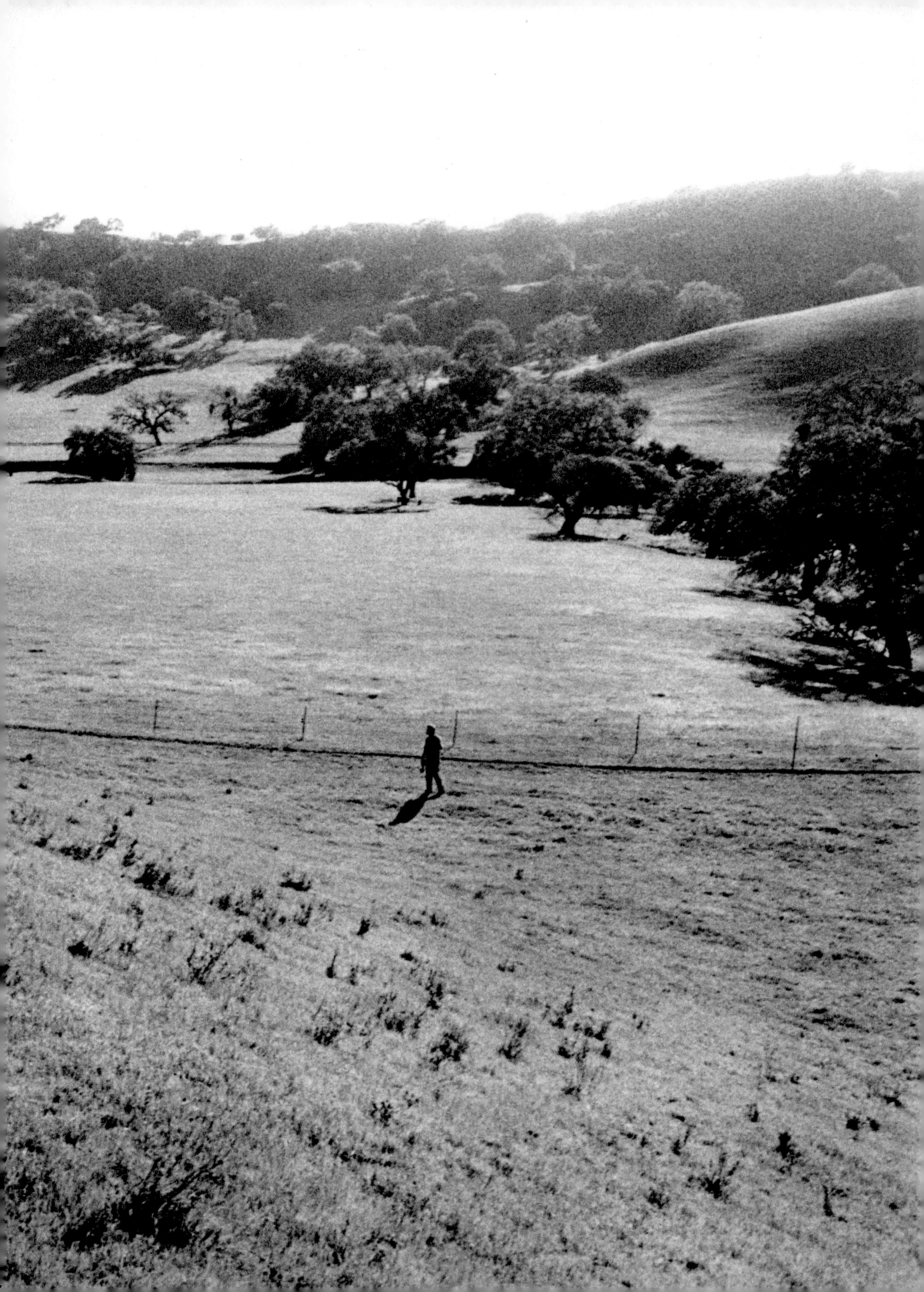

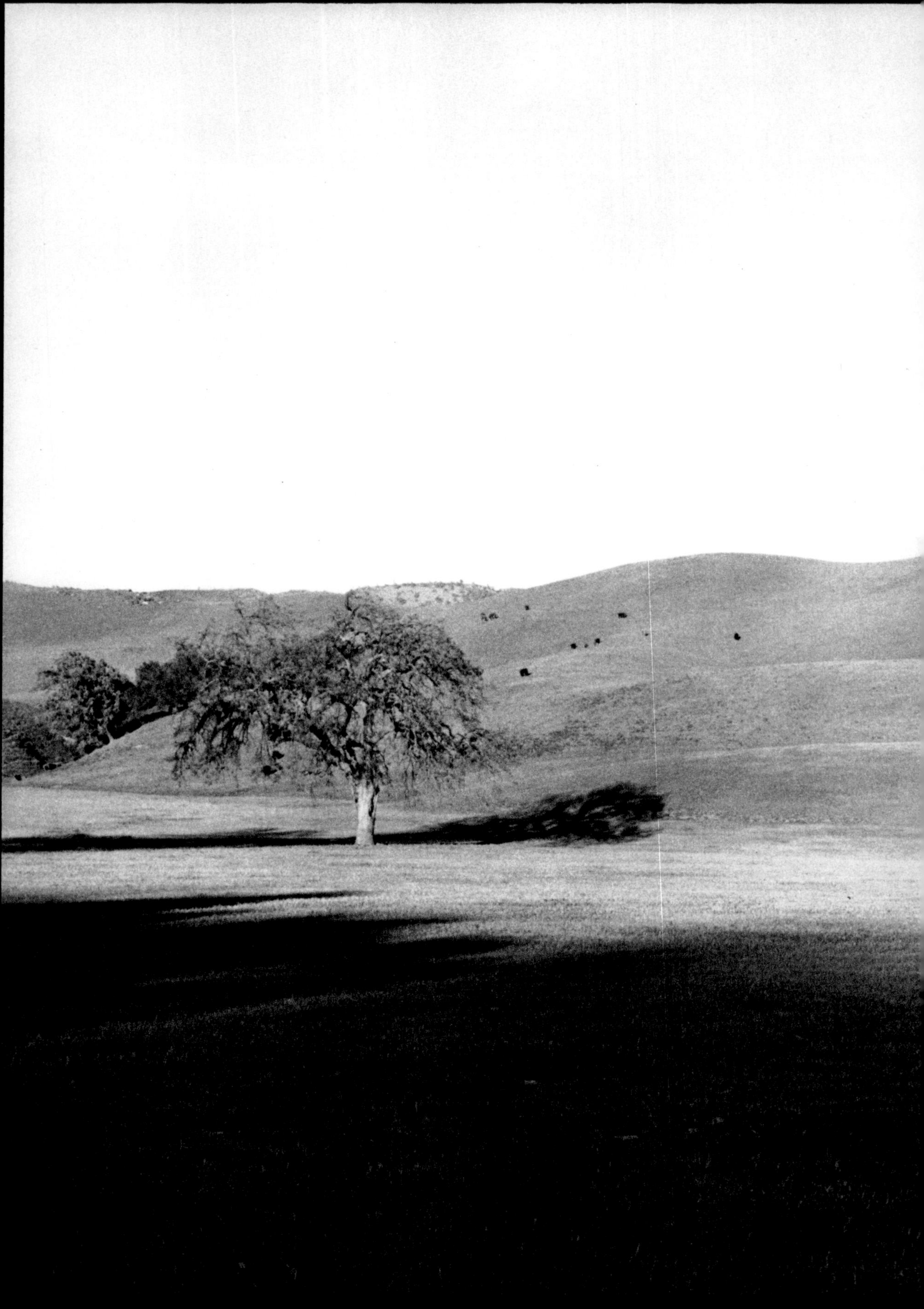

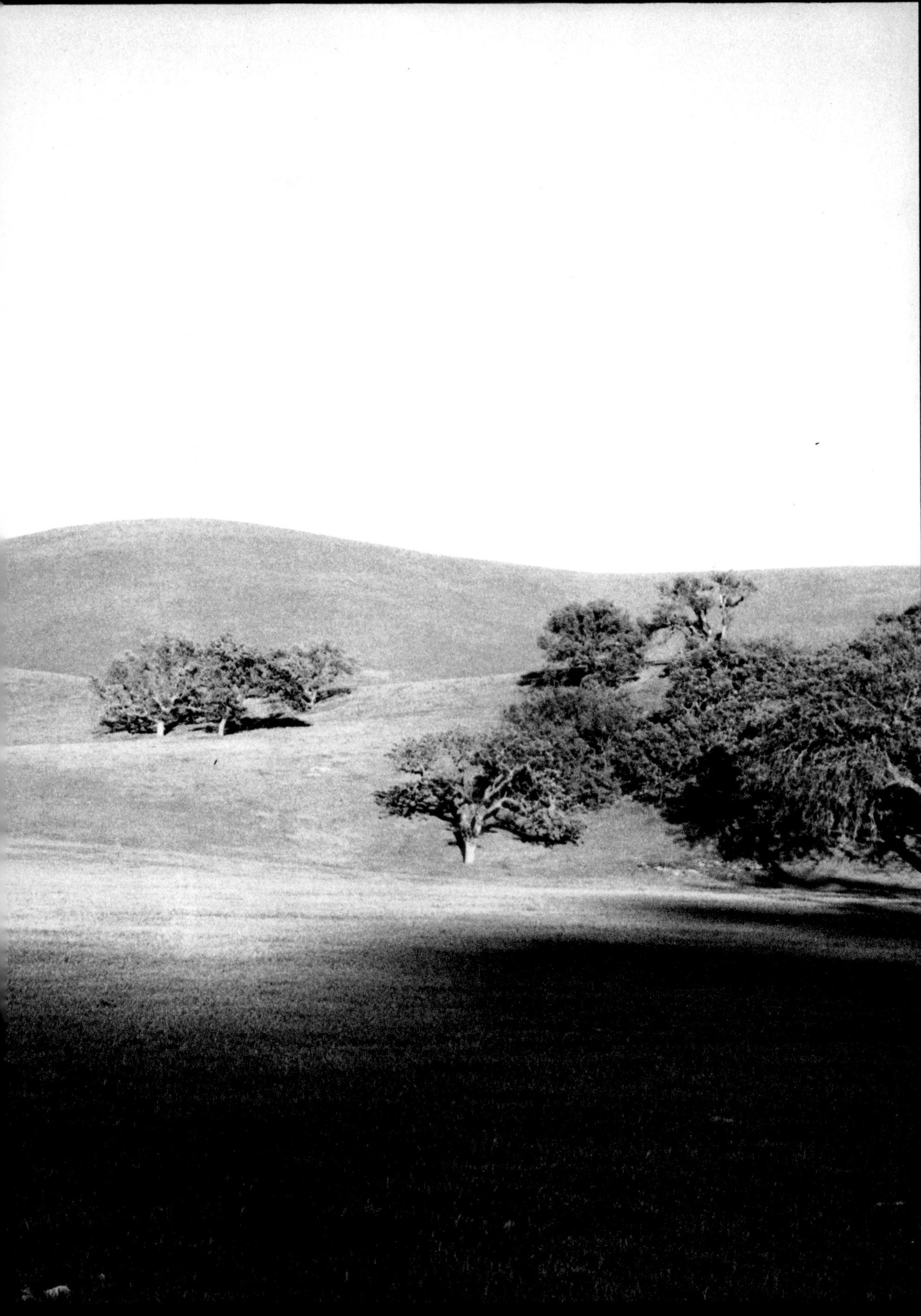

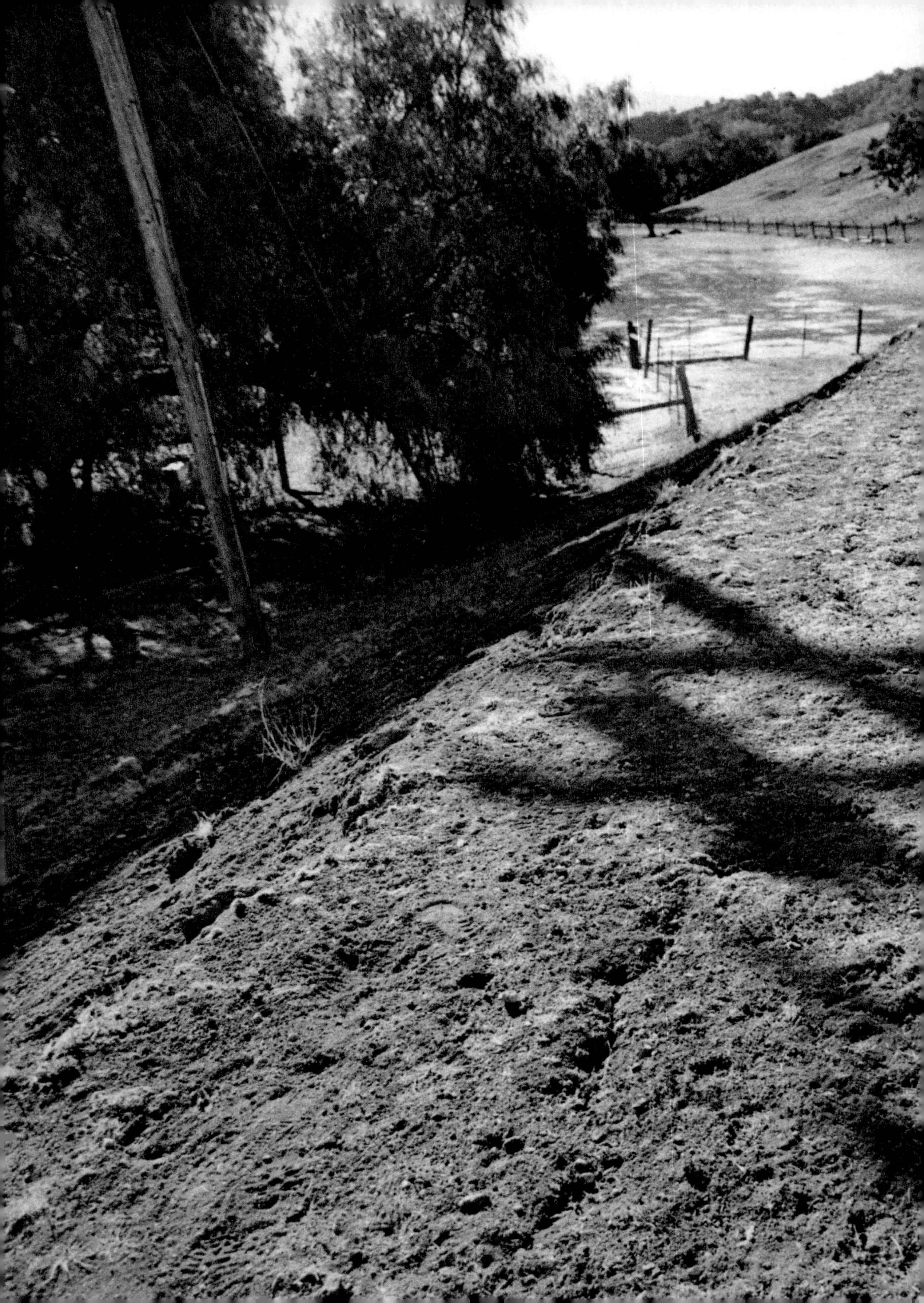

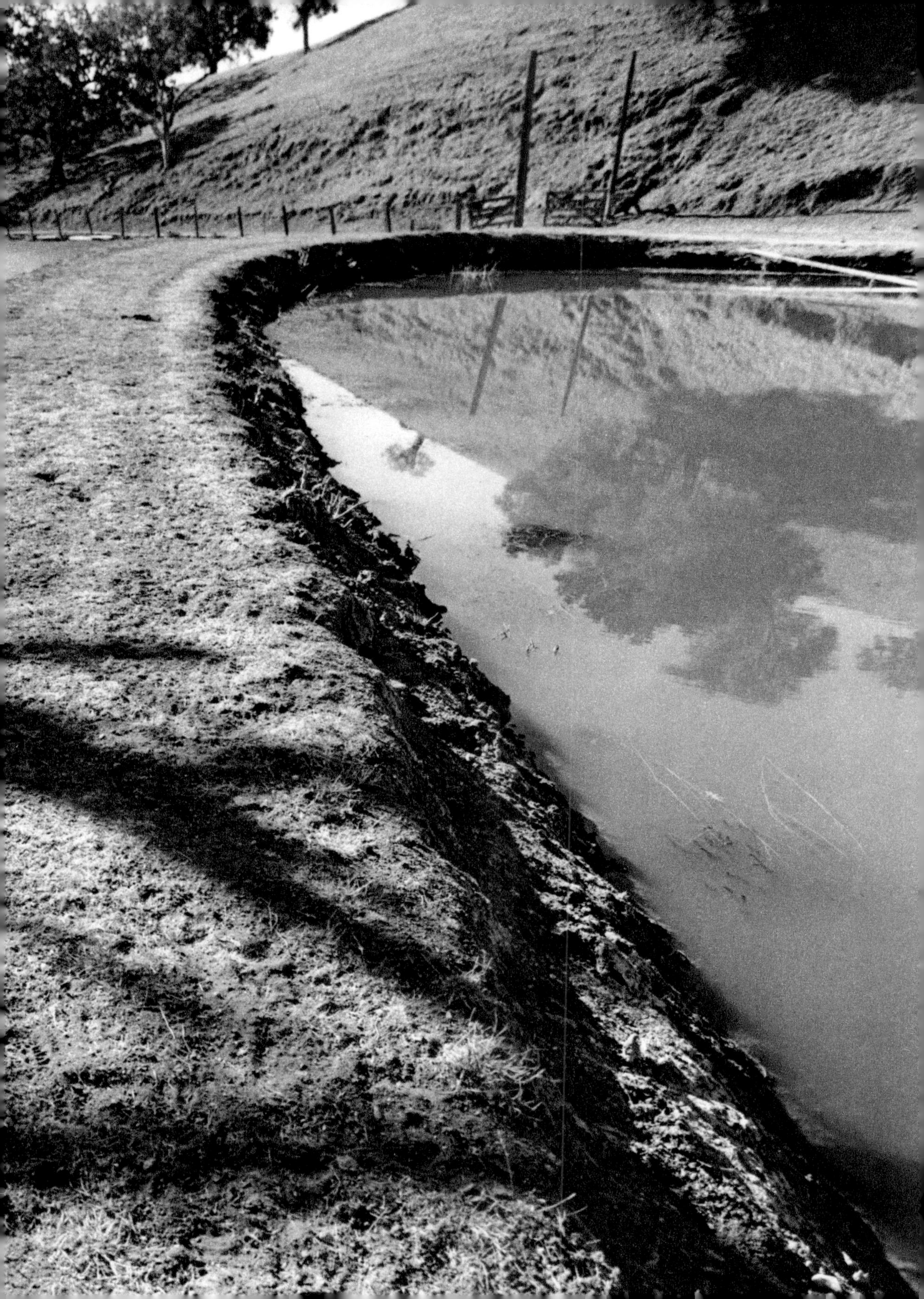

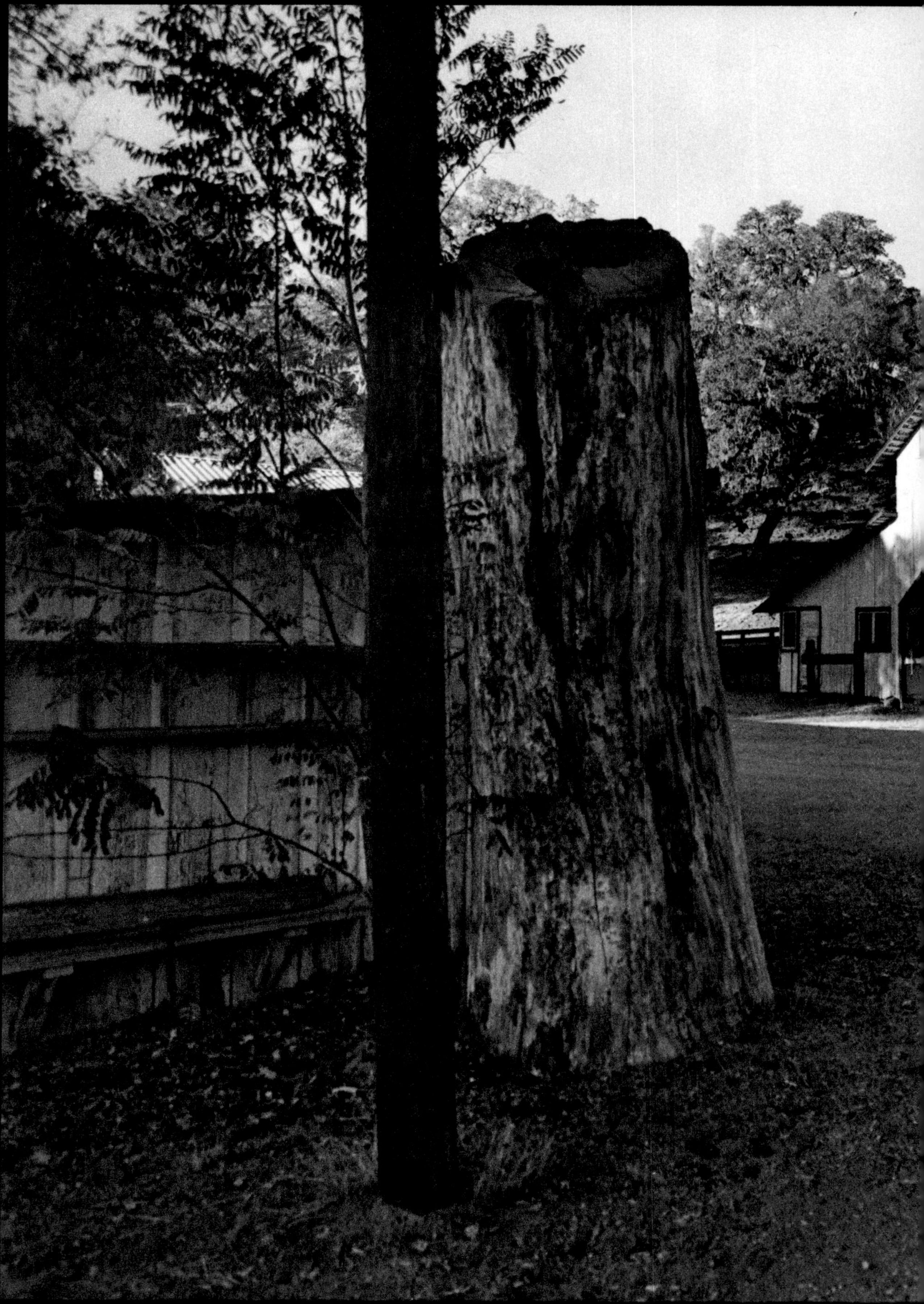

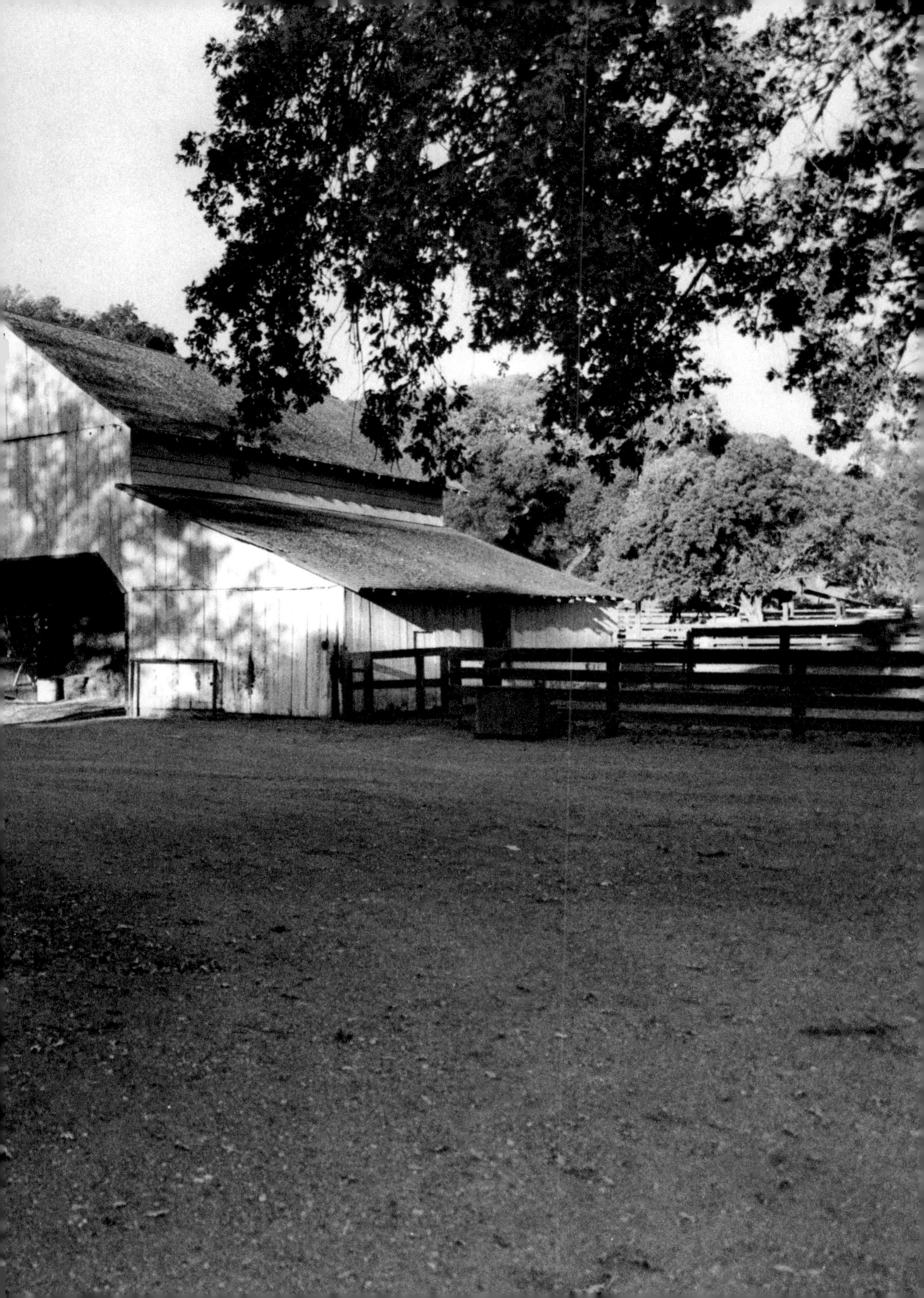

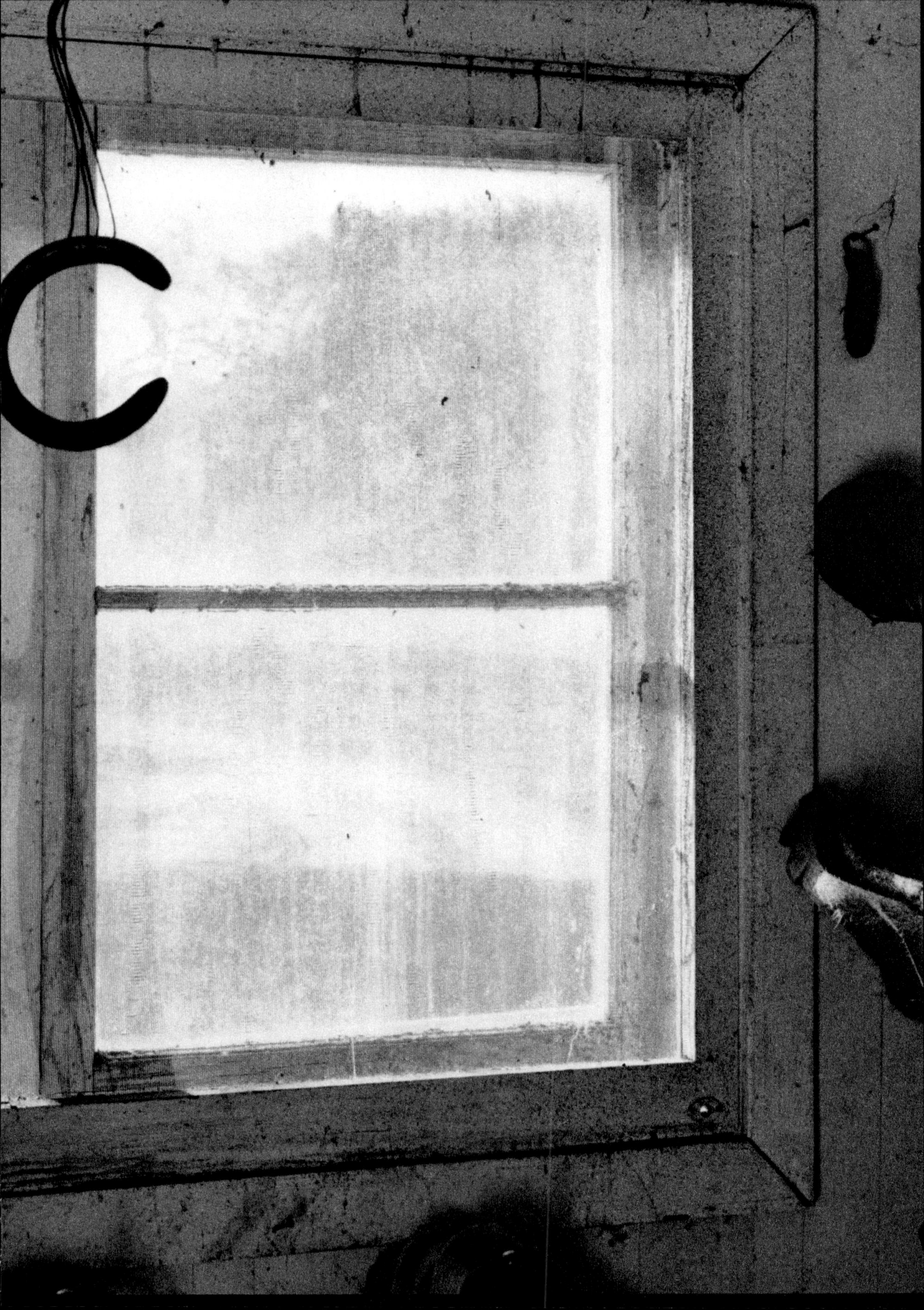

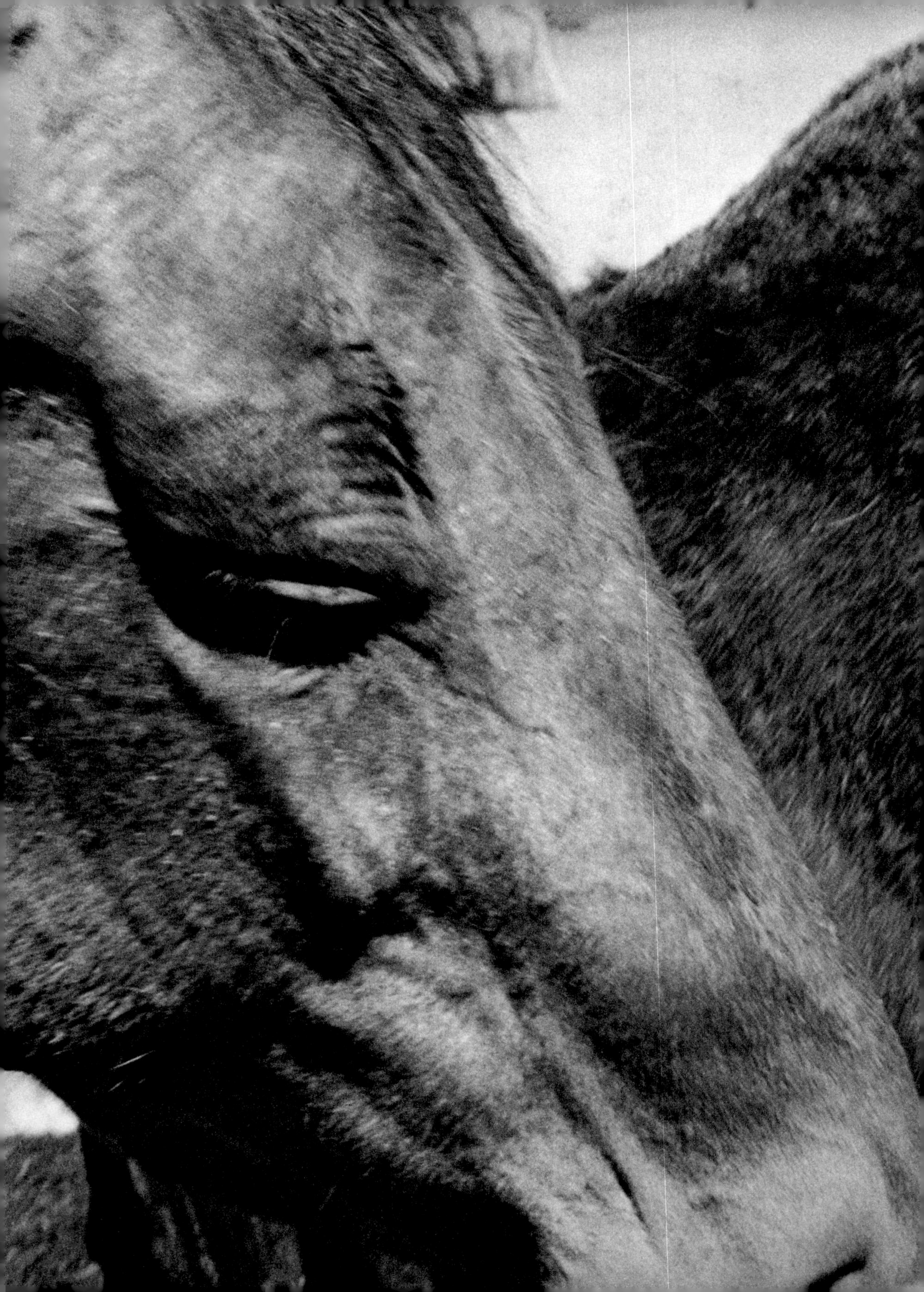

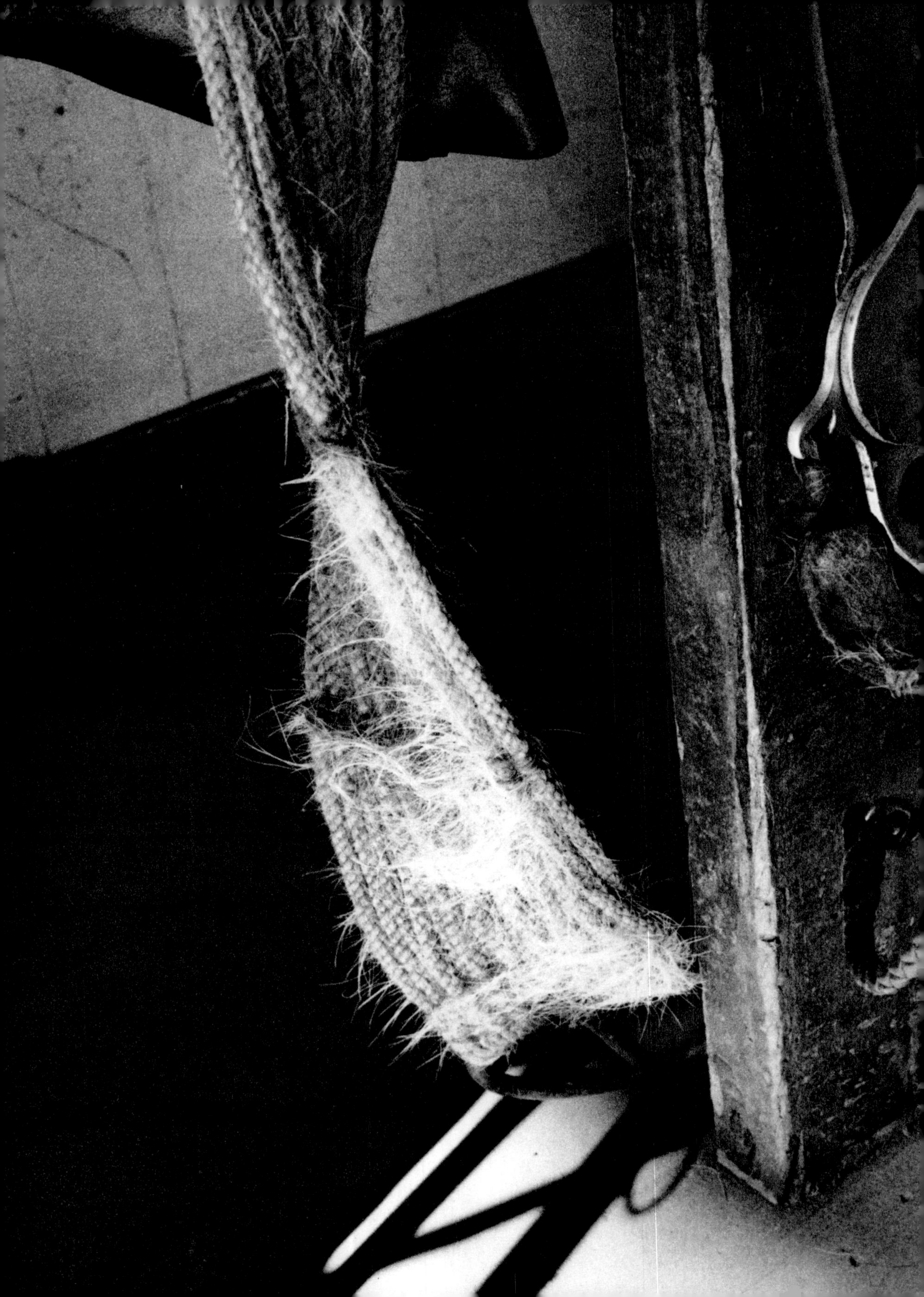

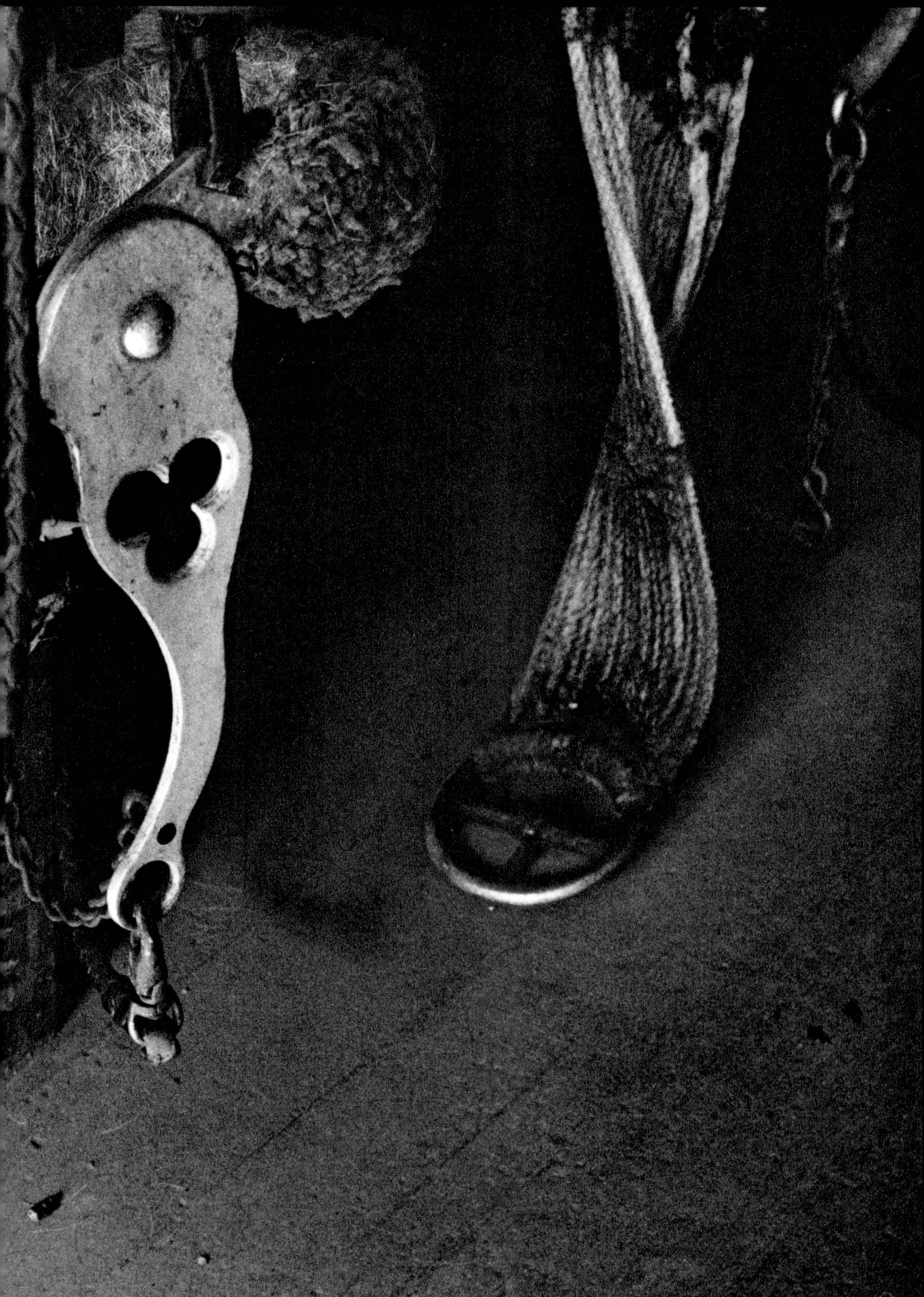

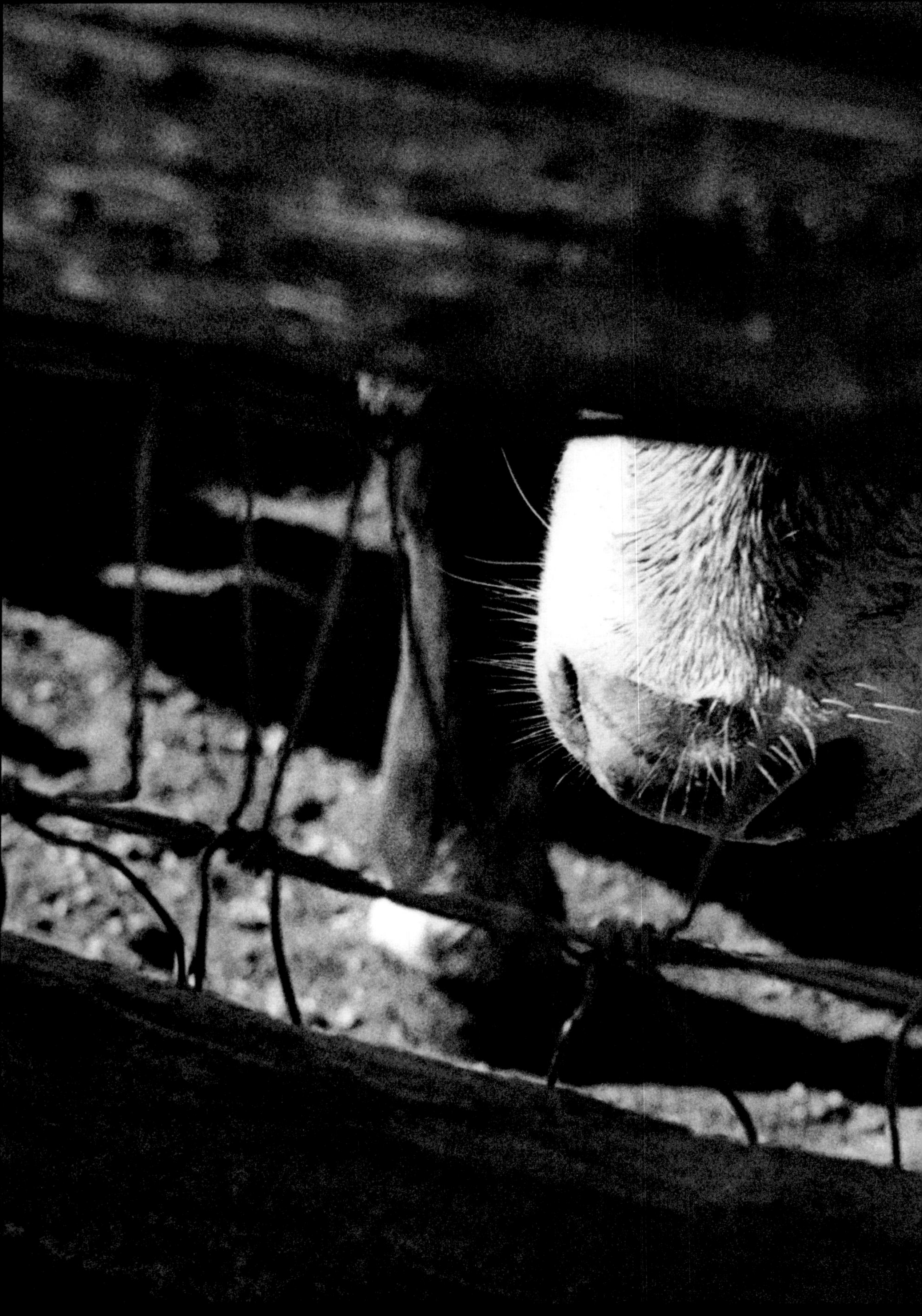

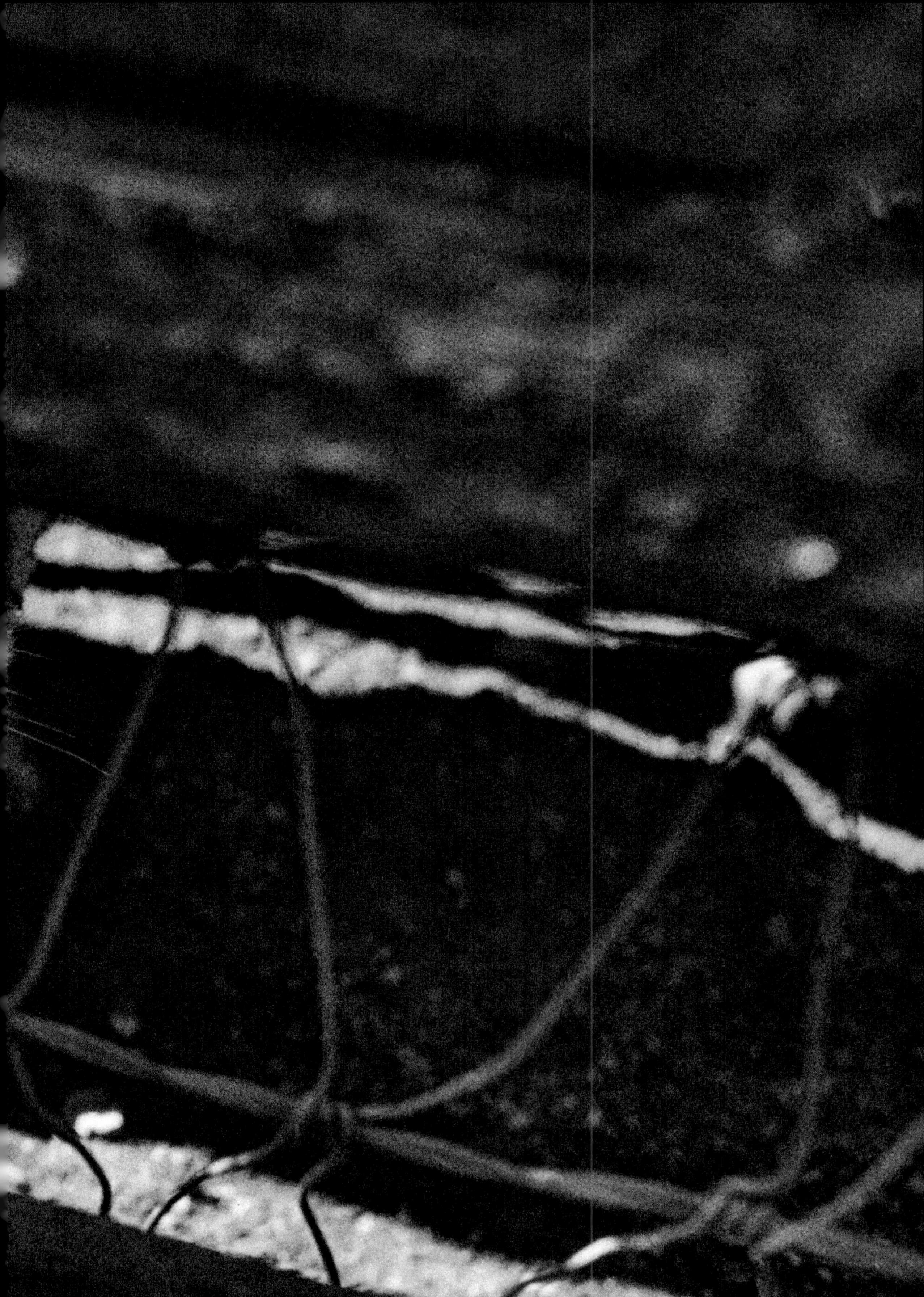

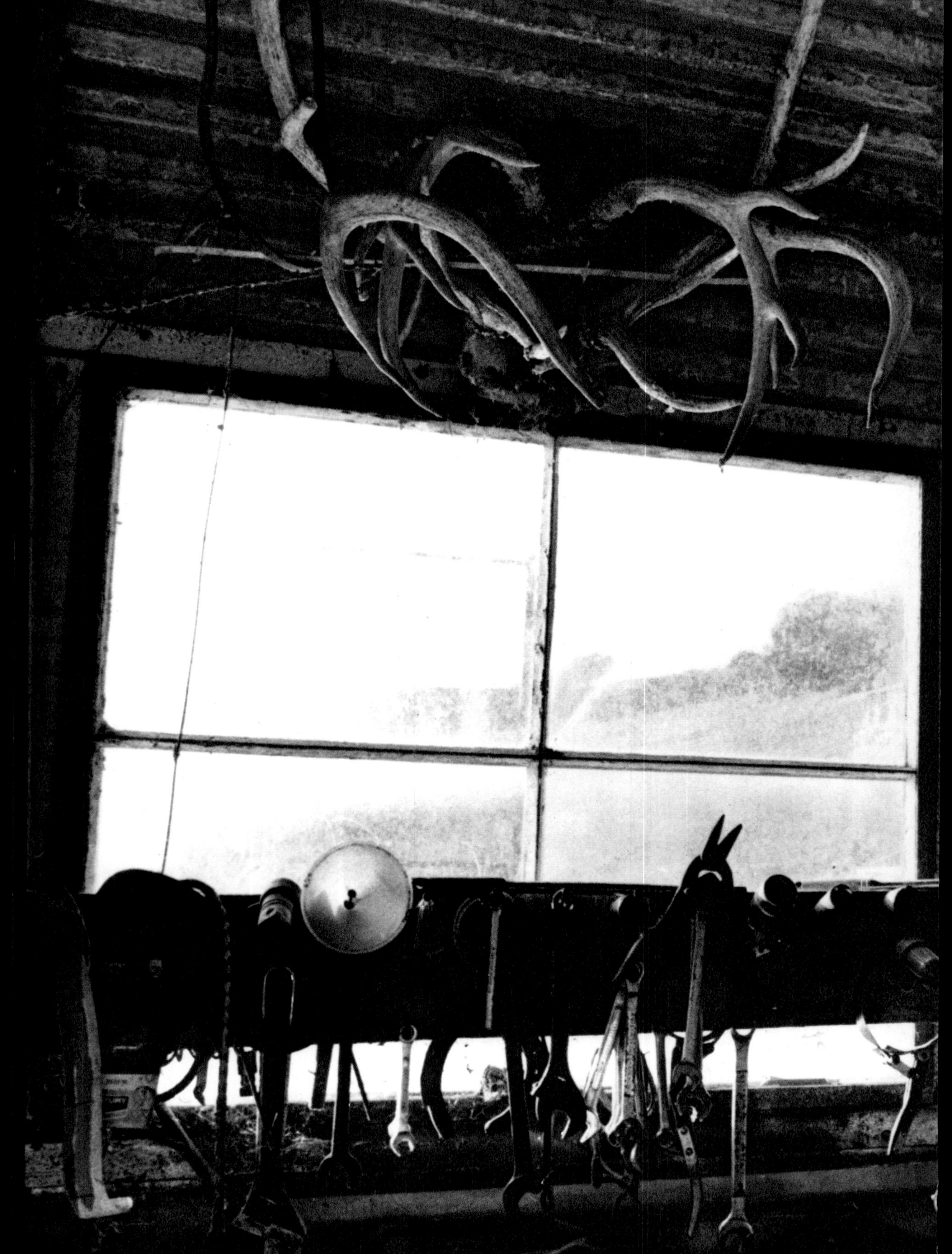

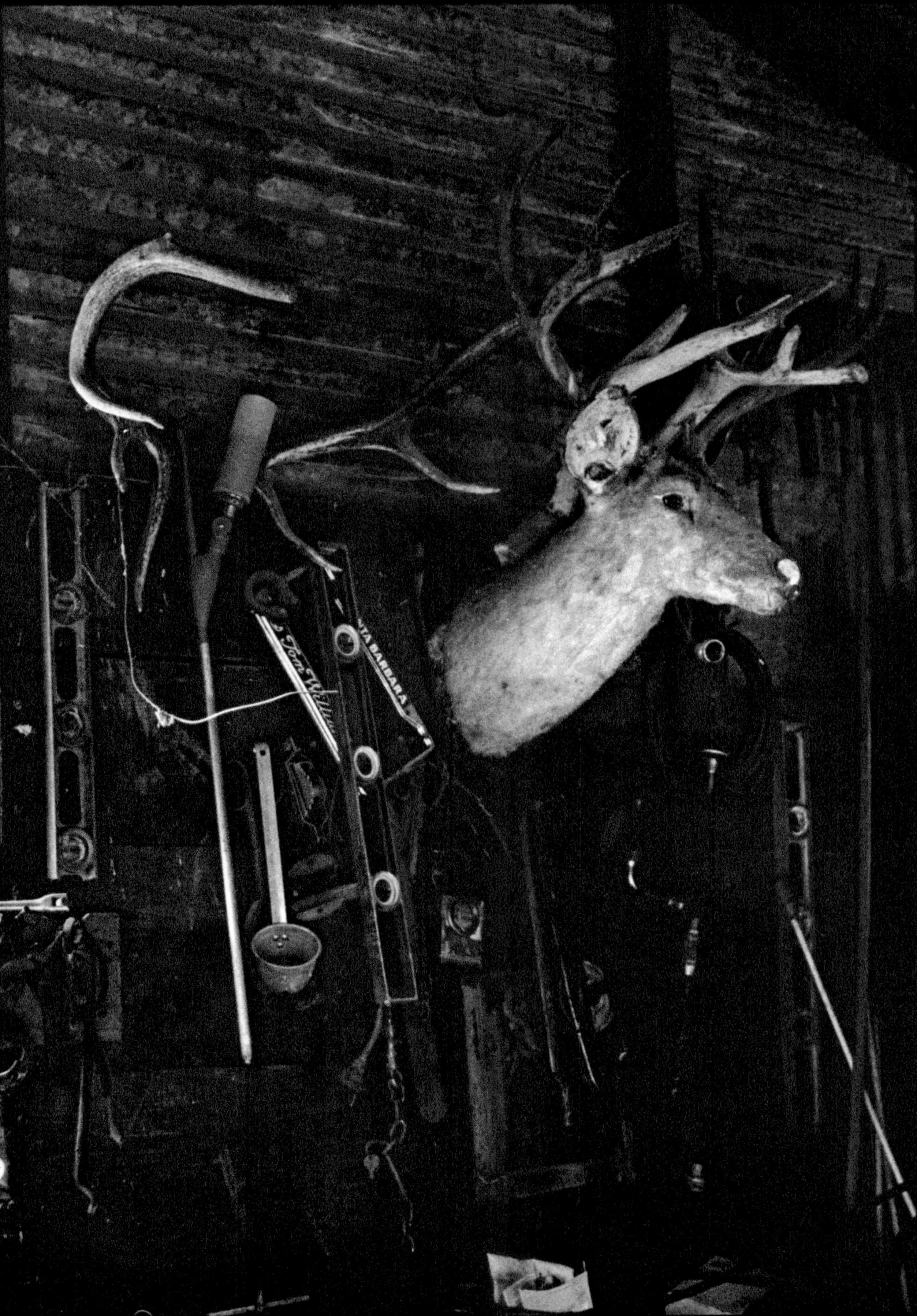

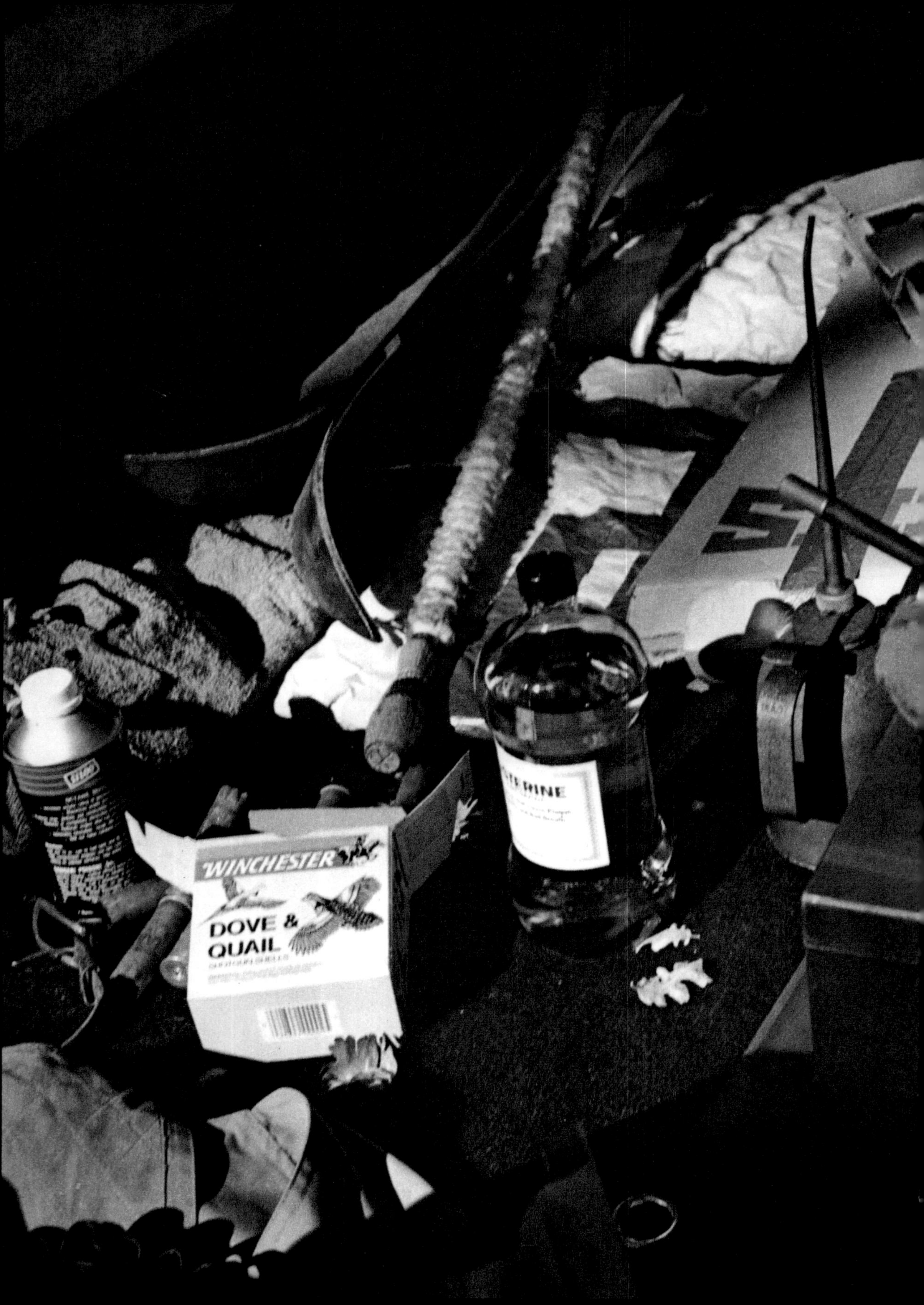
WINCHESTER
DOVE &
QUAIL

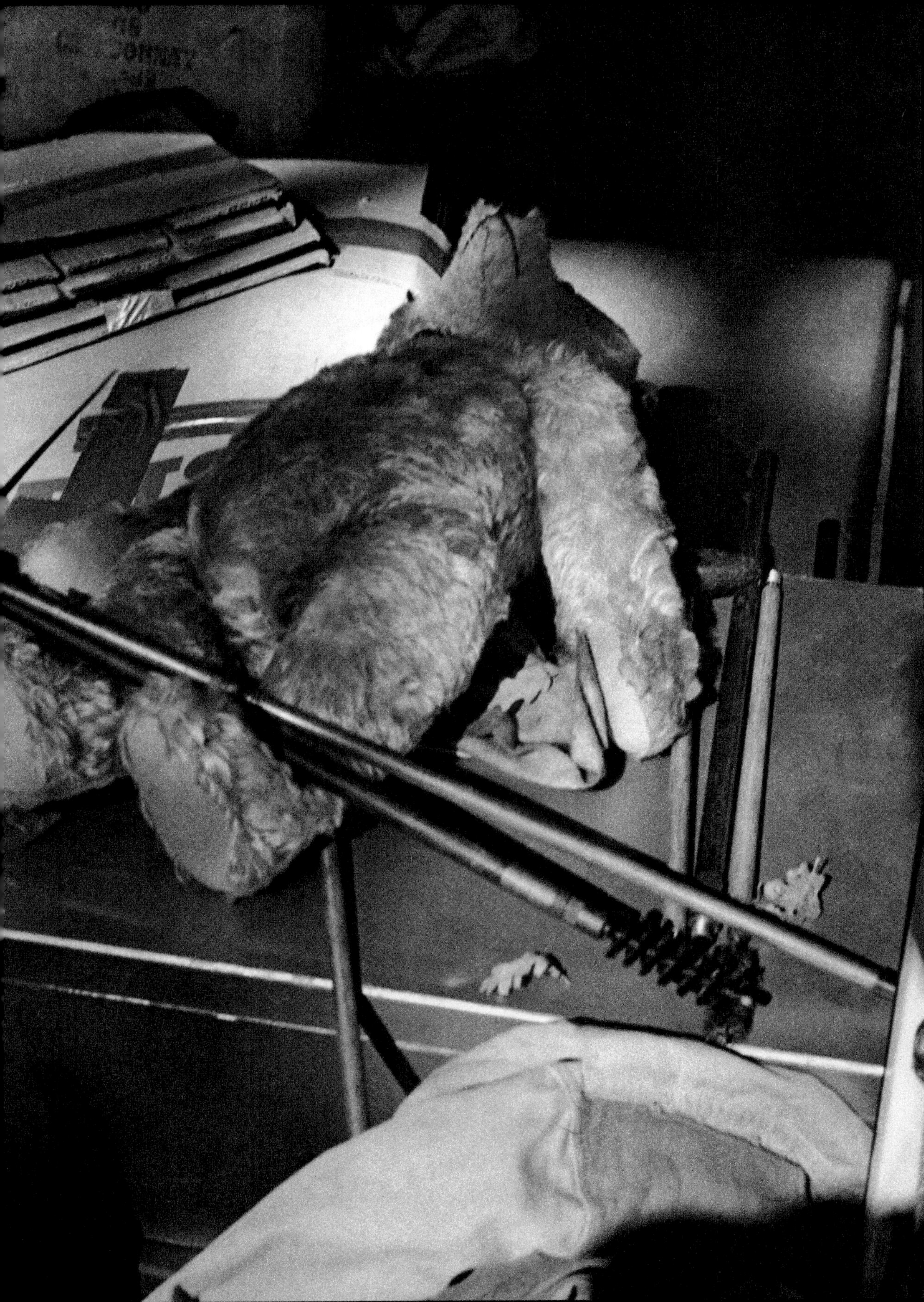

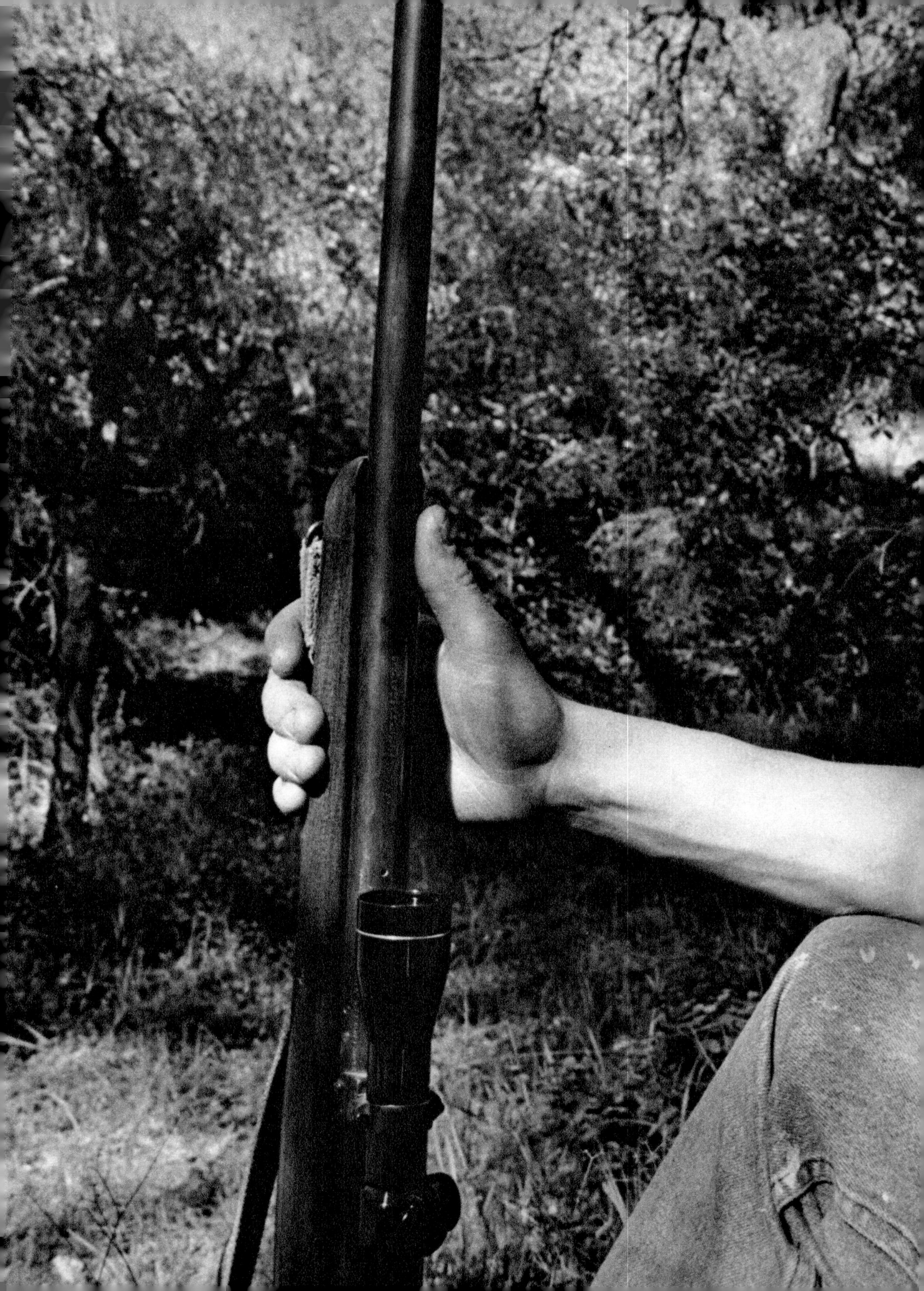

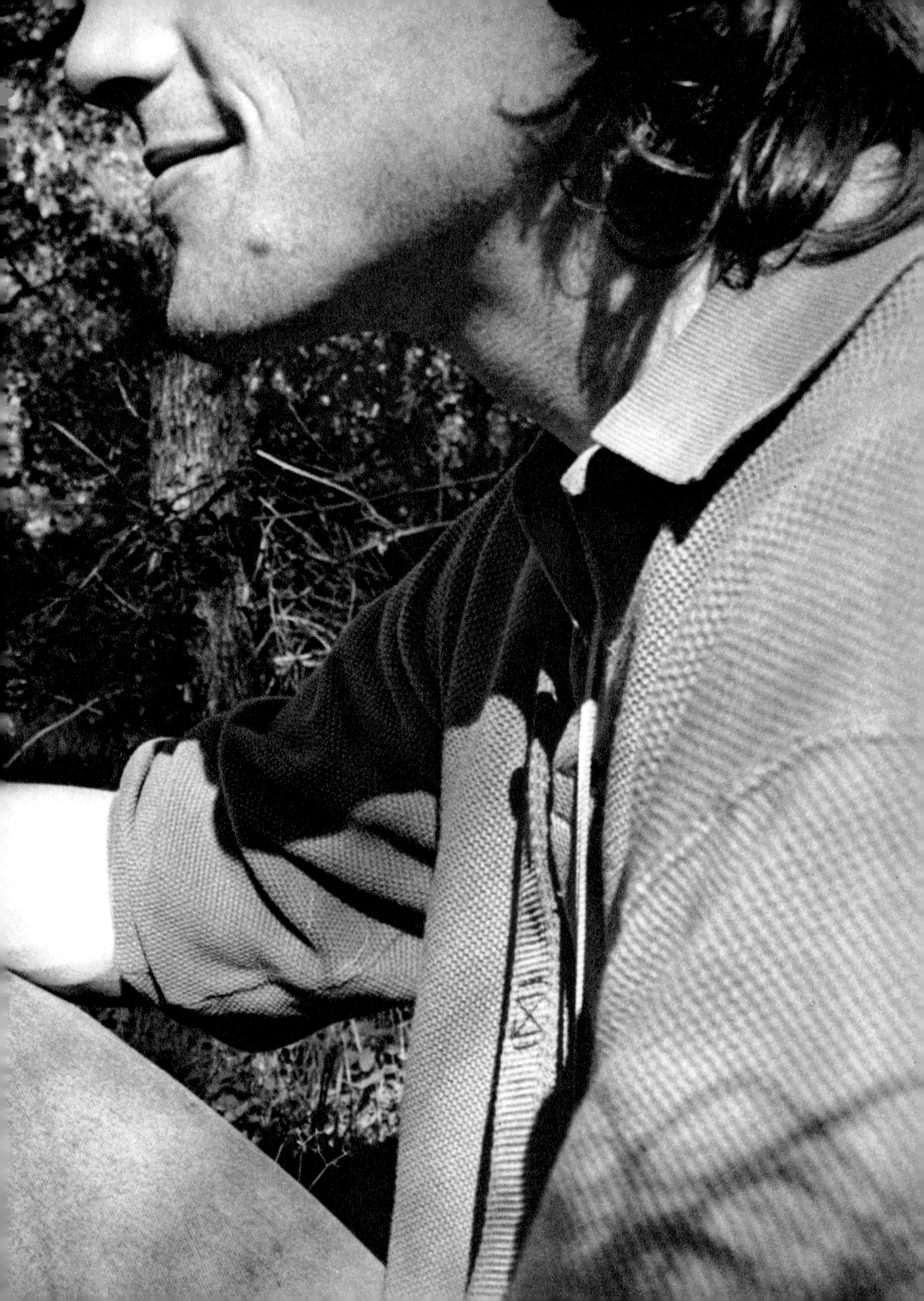

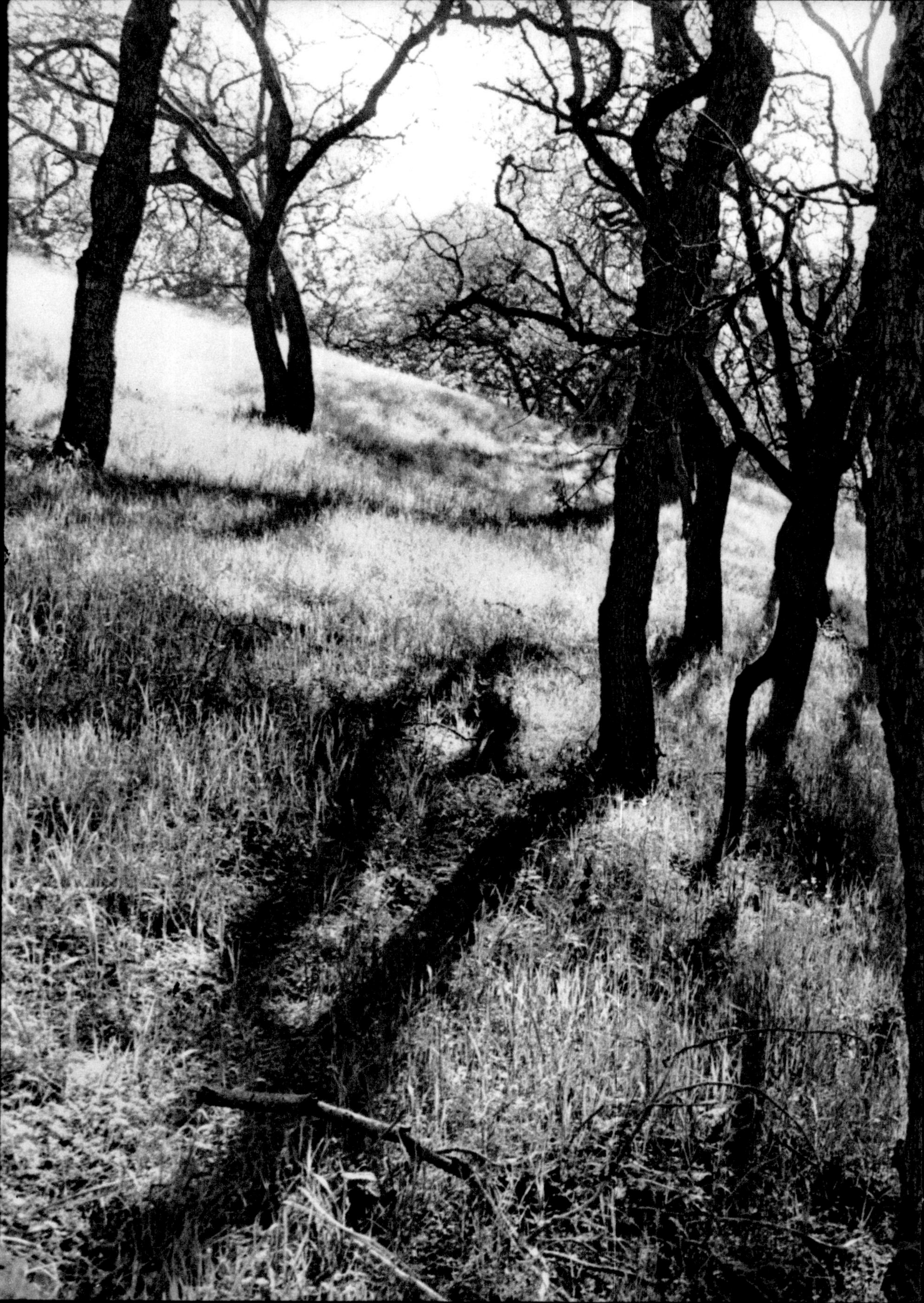

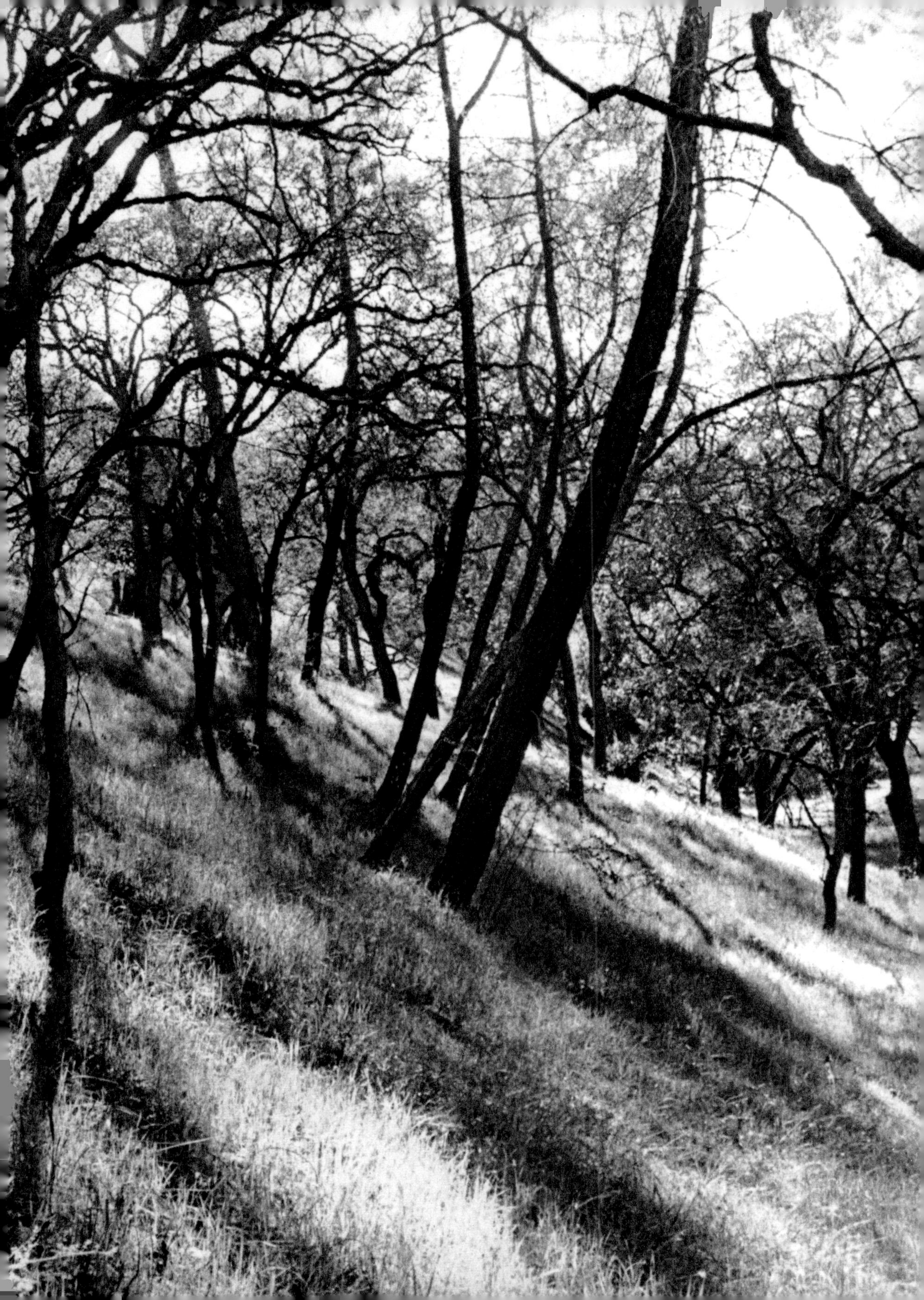

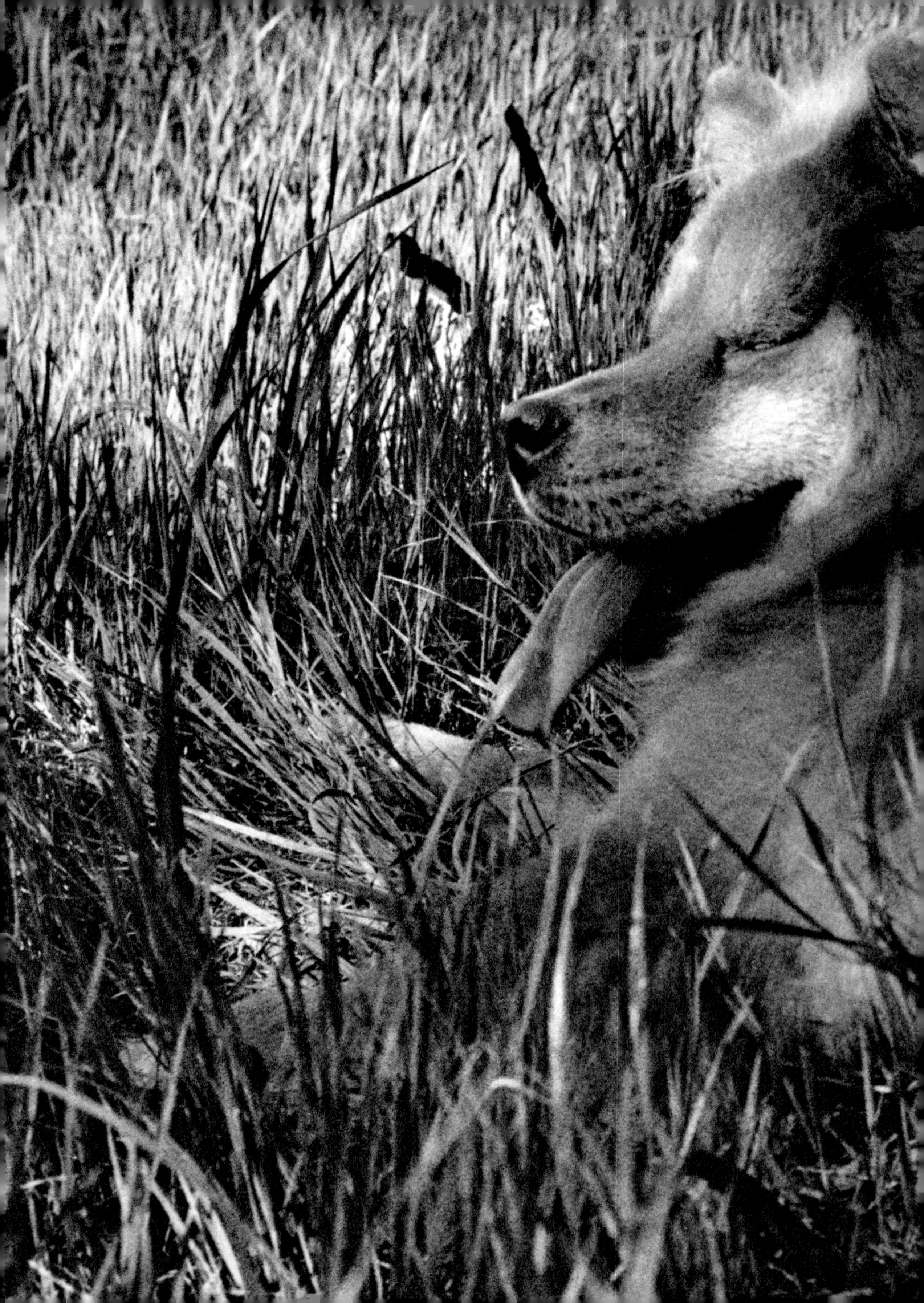

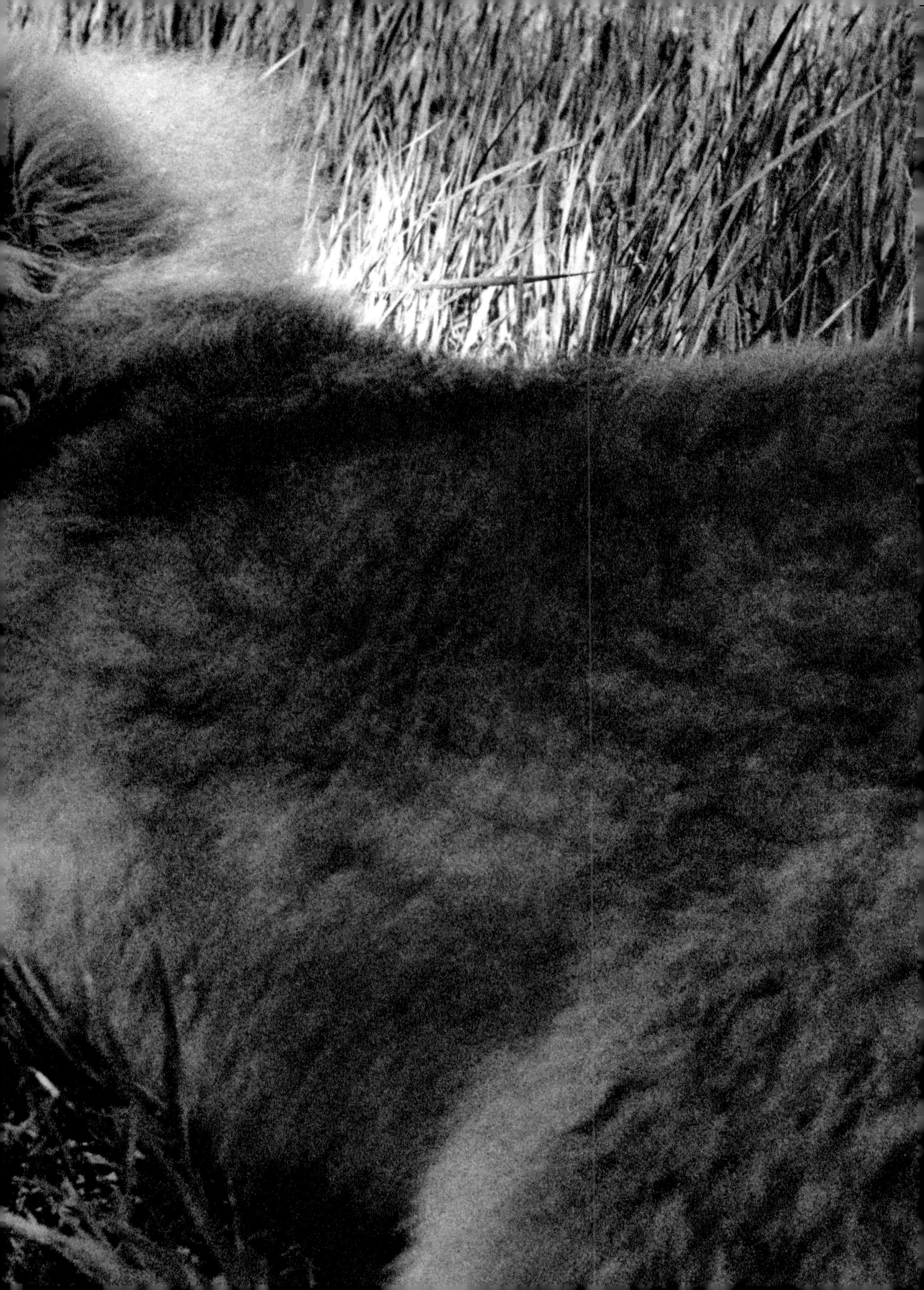

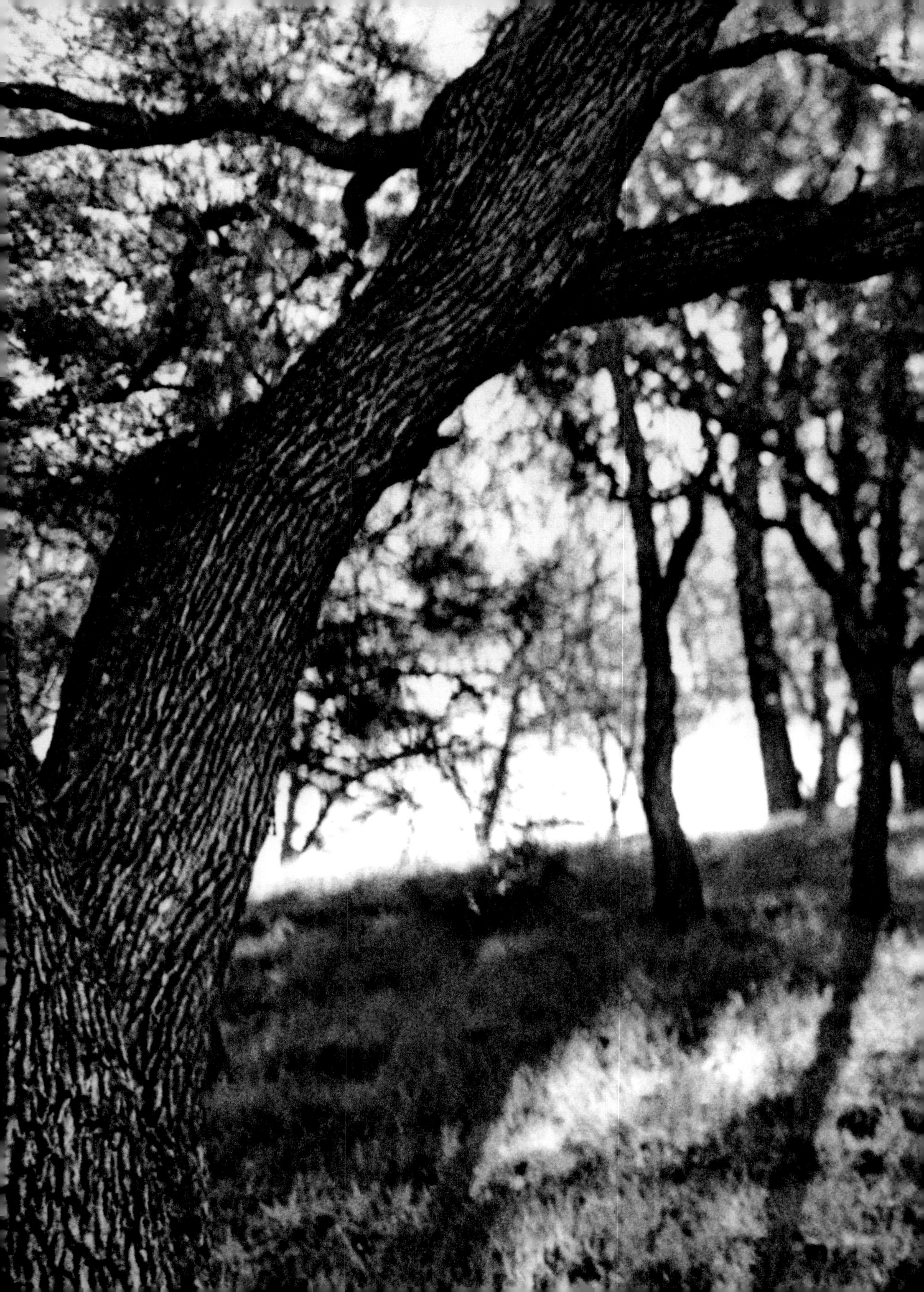

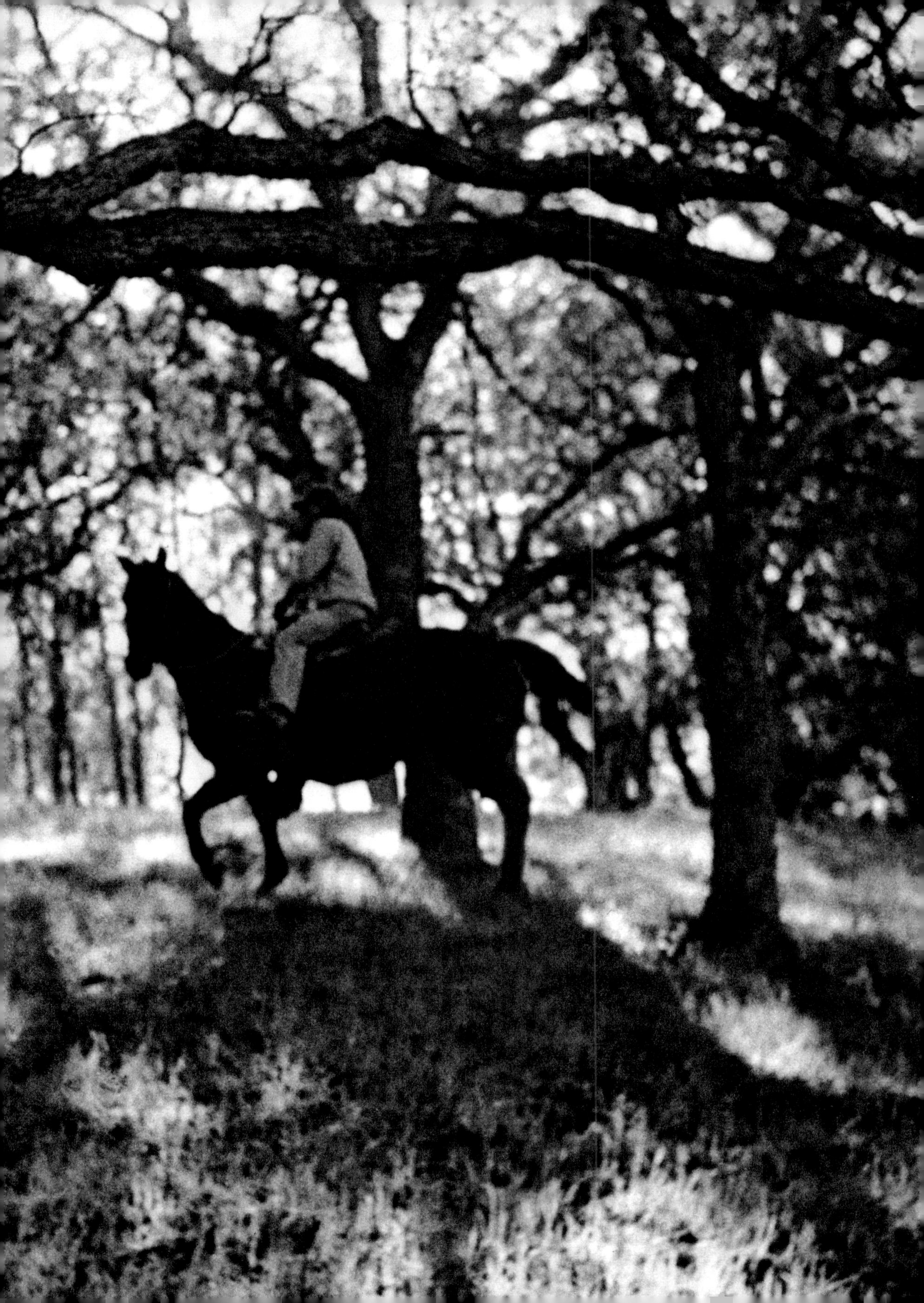

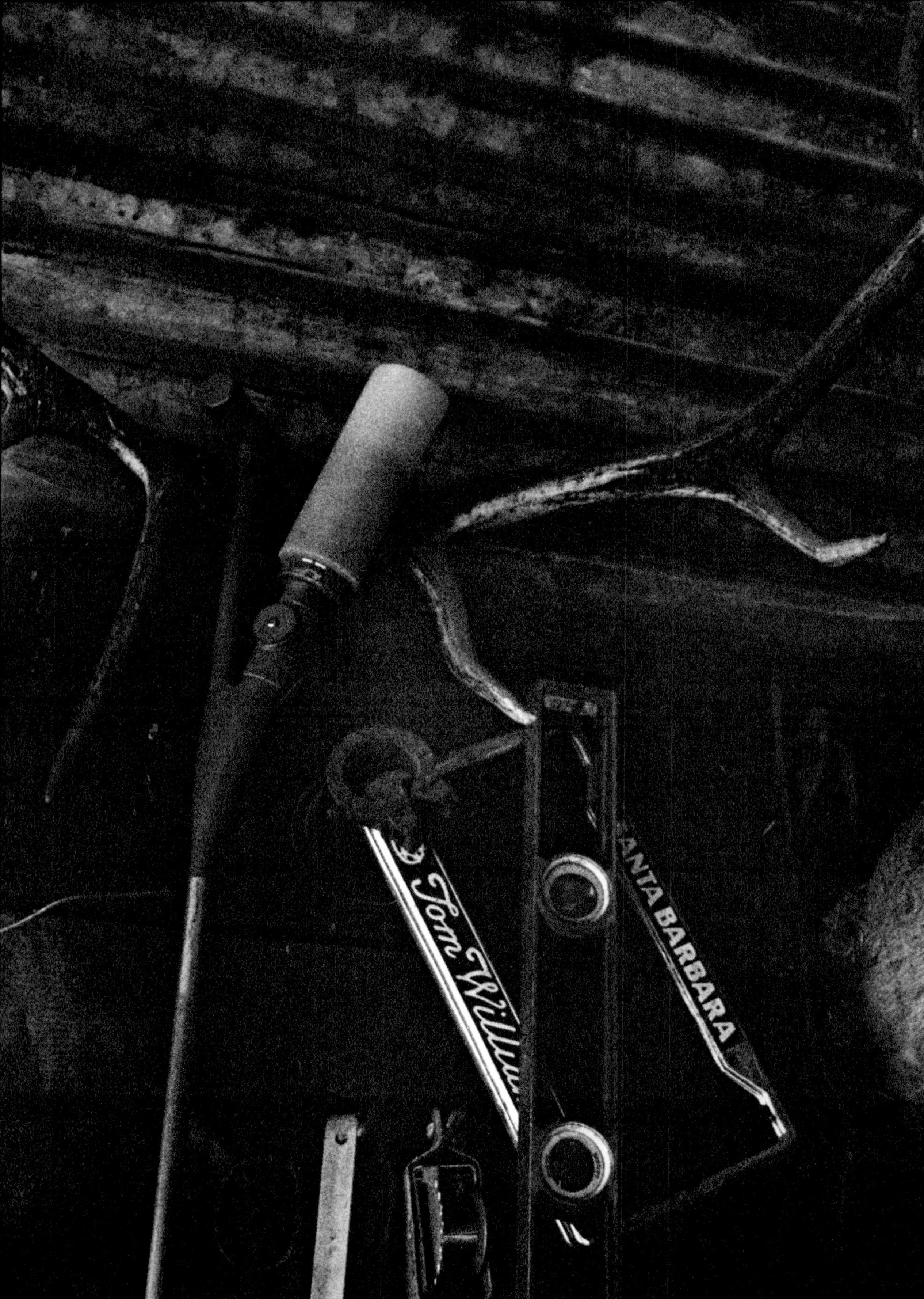

Tom Will
ANTA BARBARA

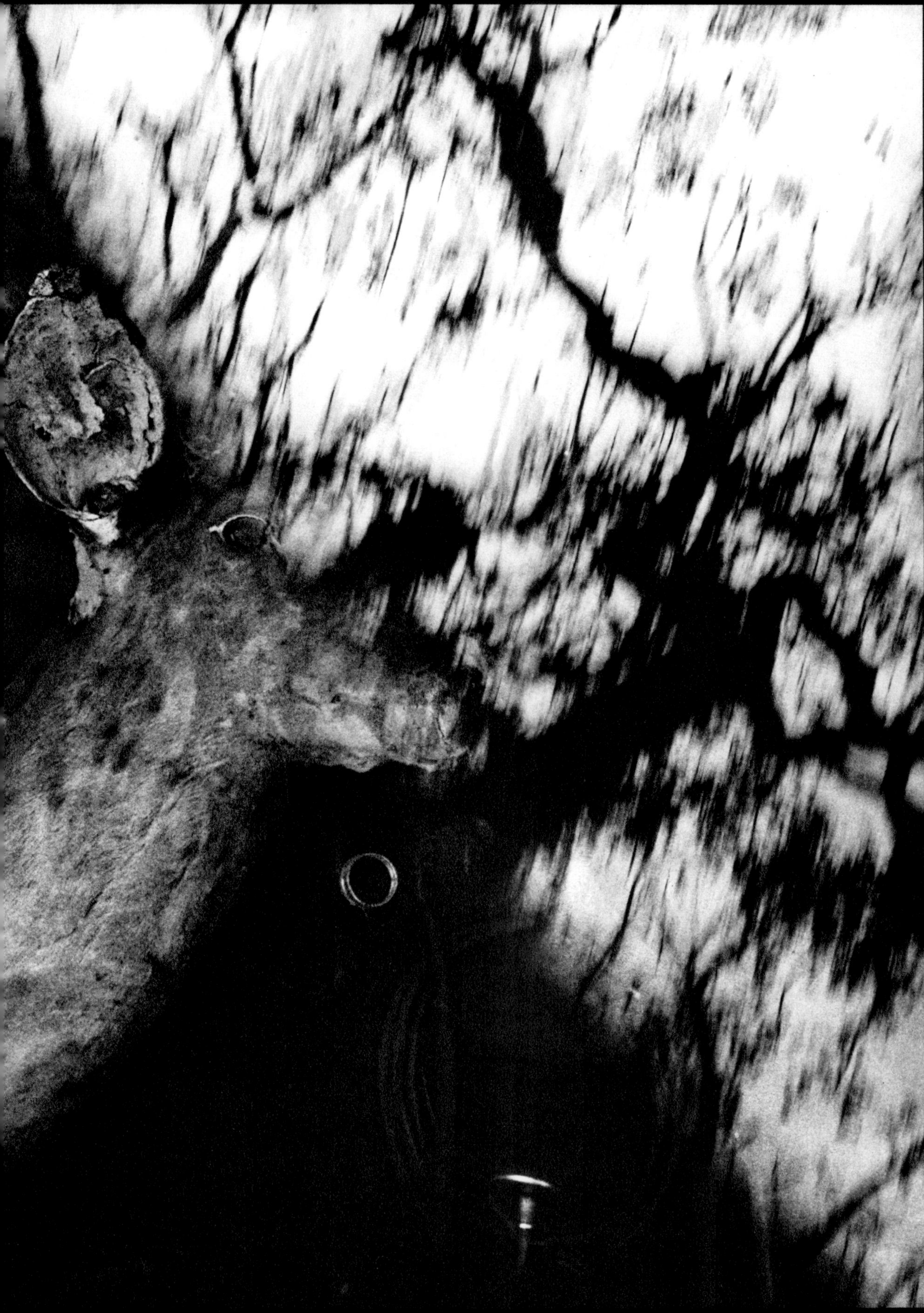

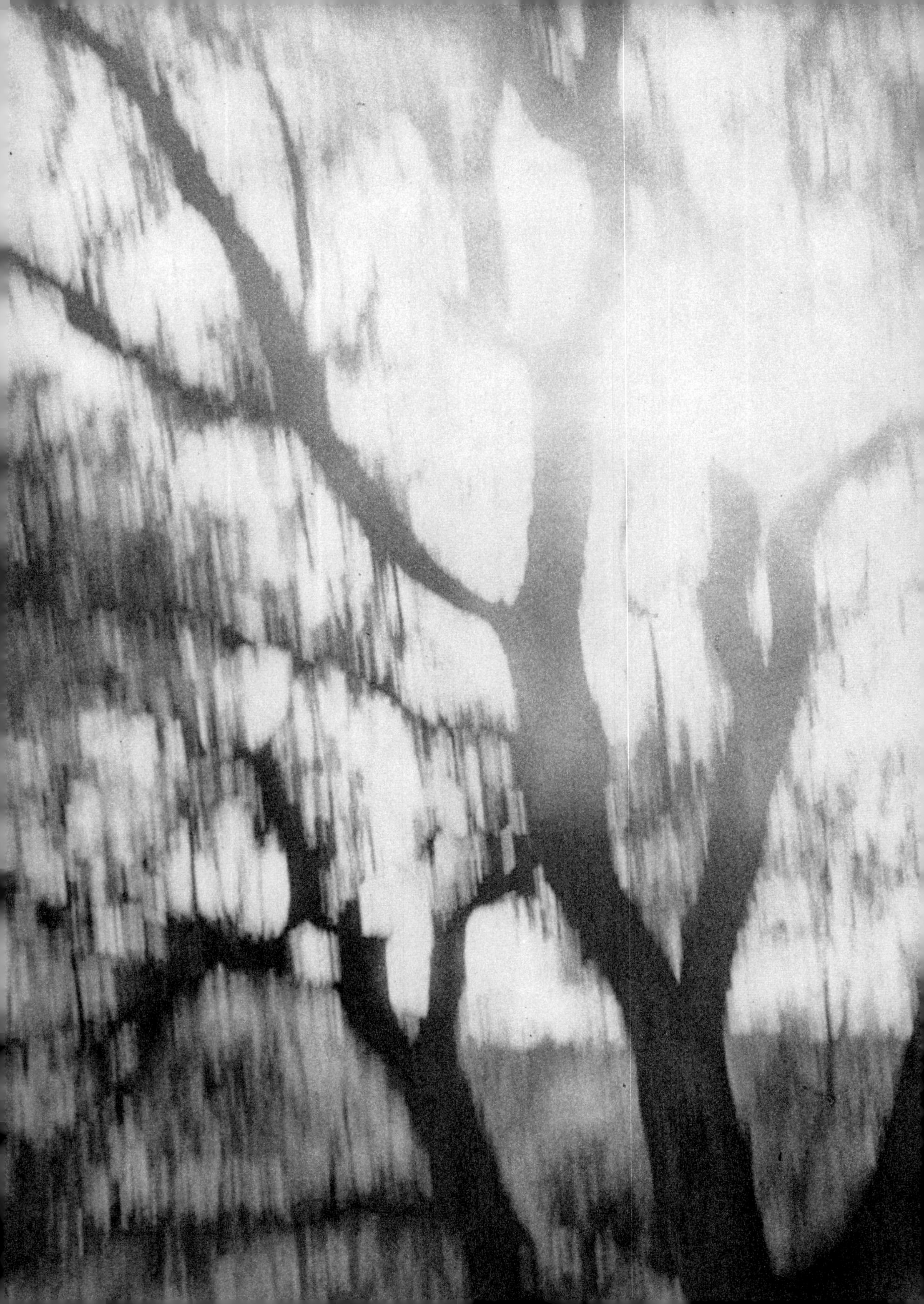

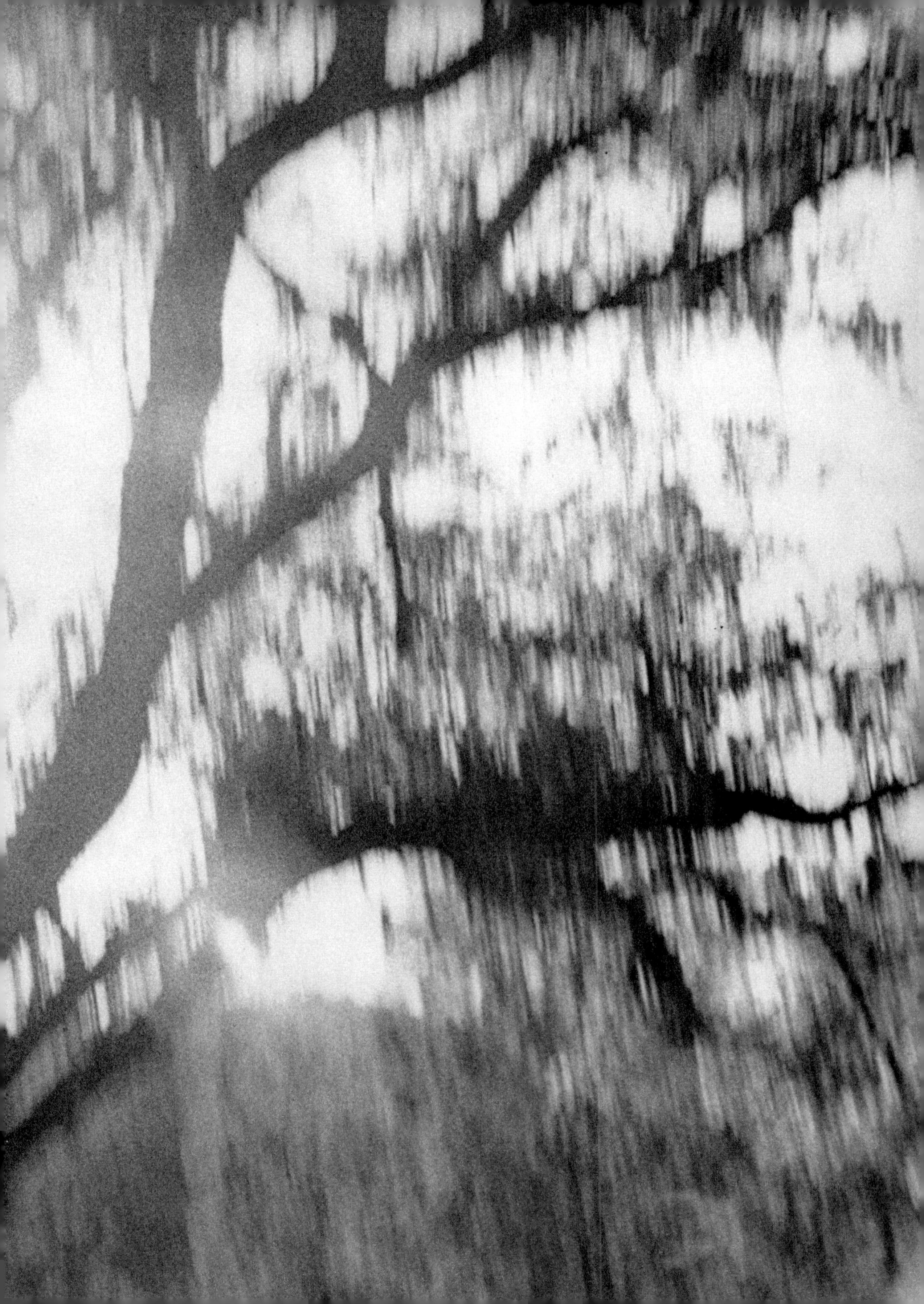

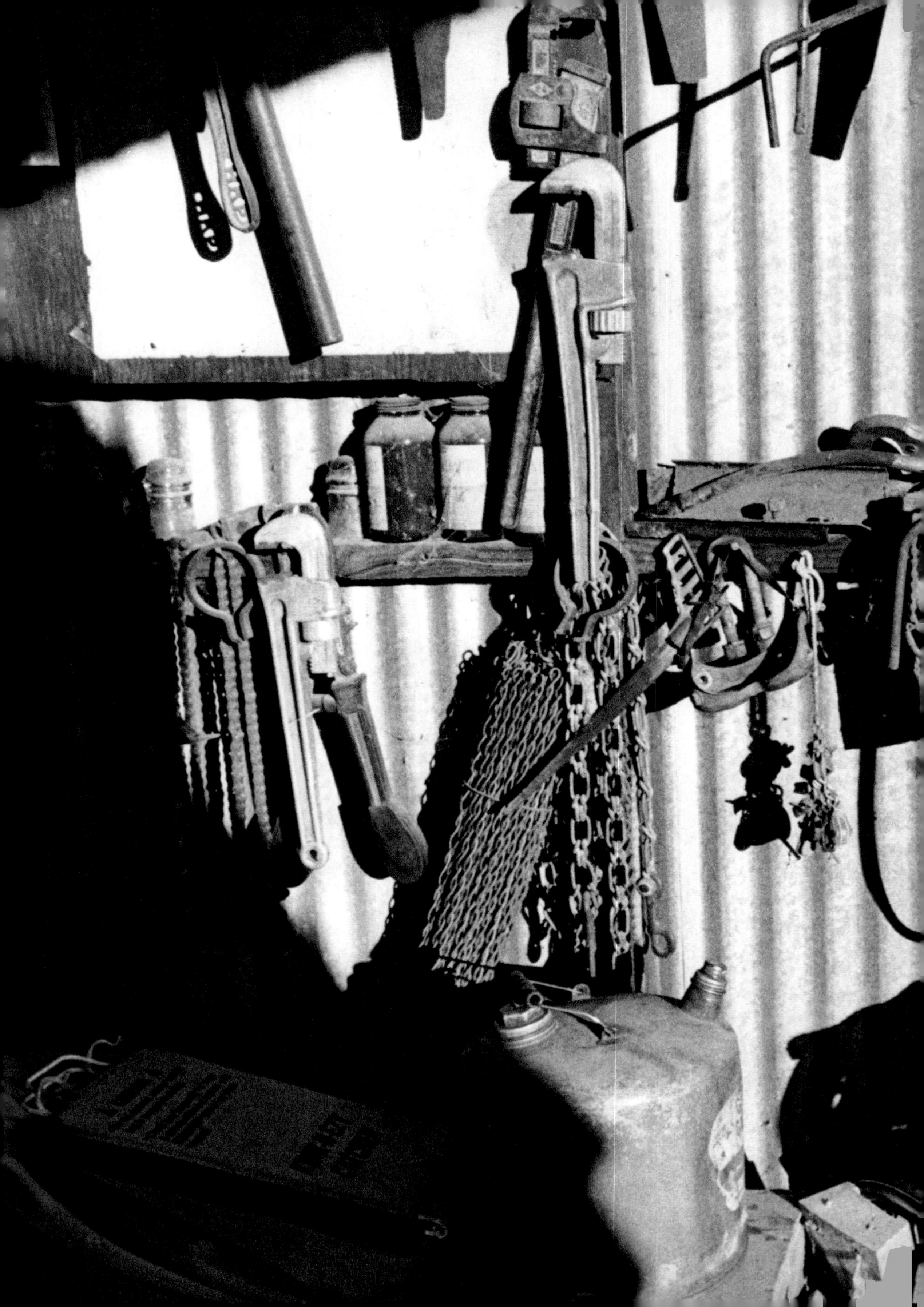

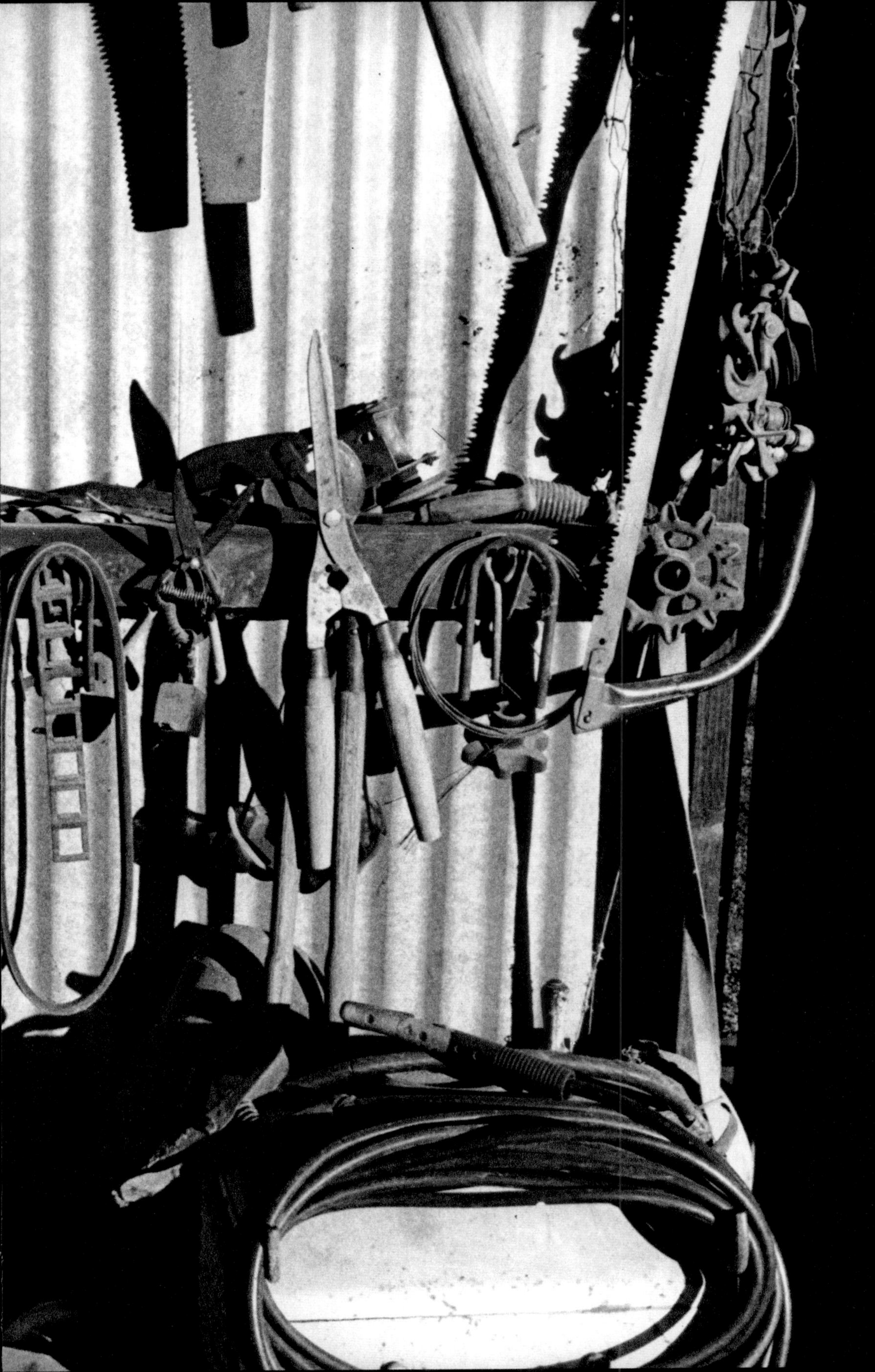

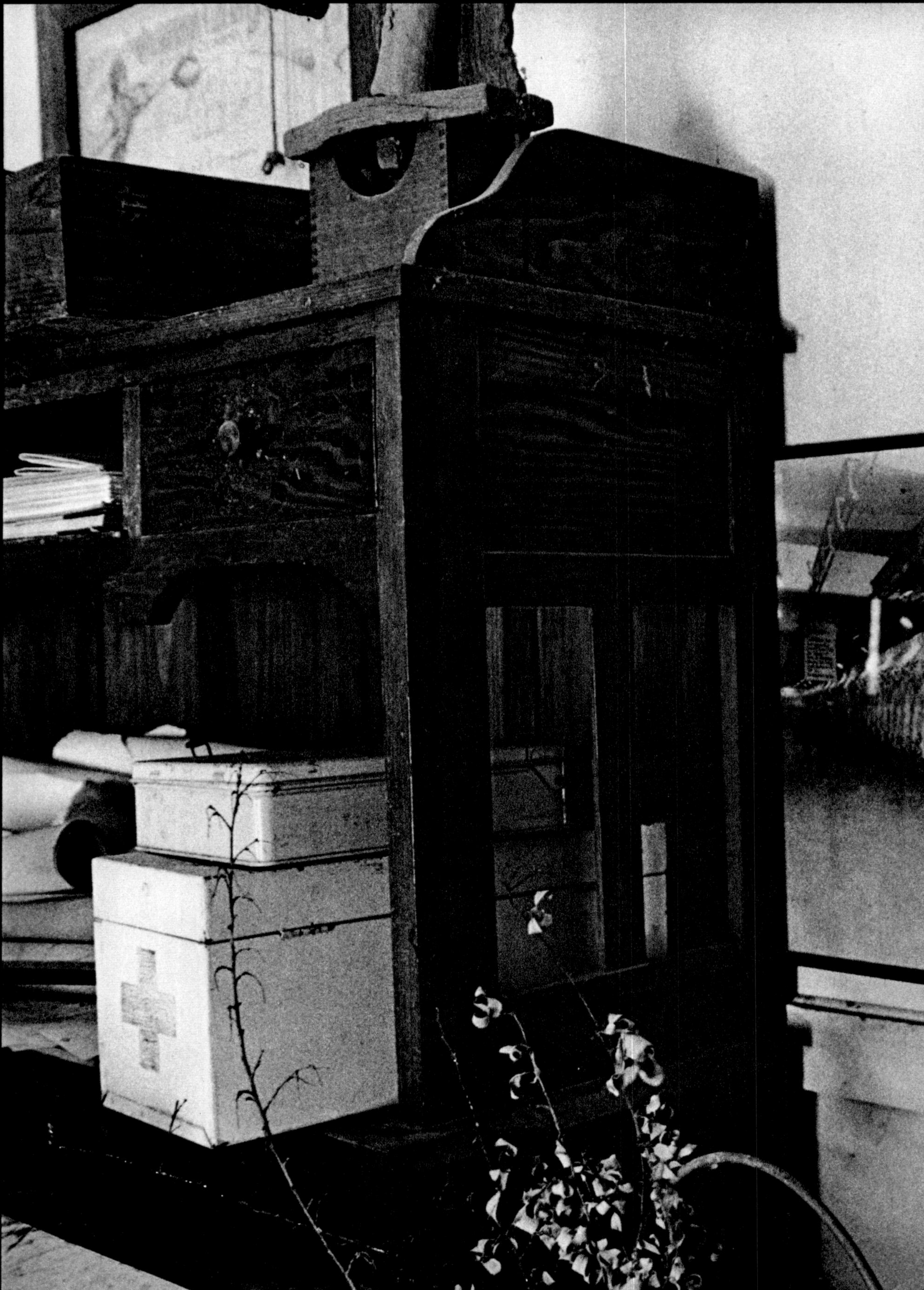

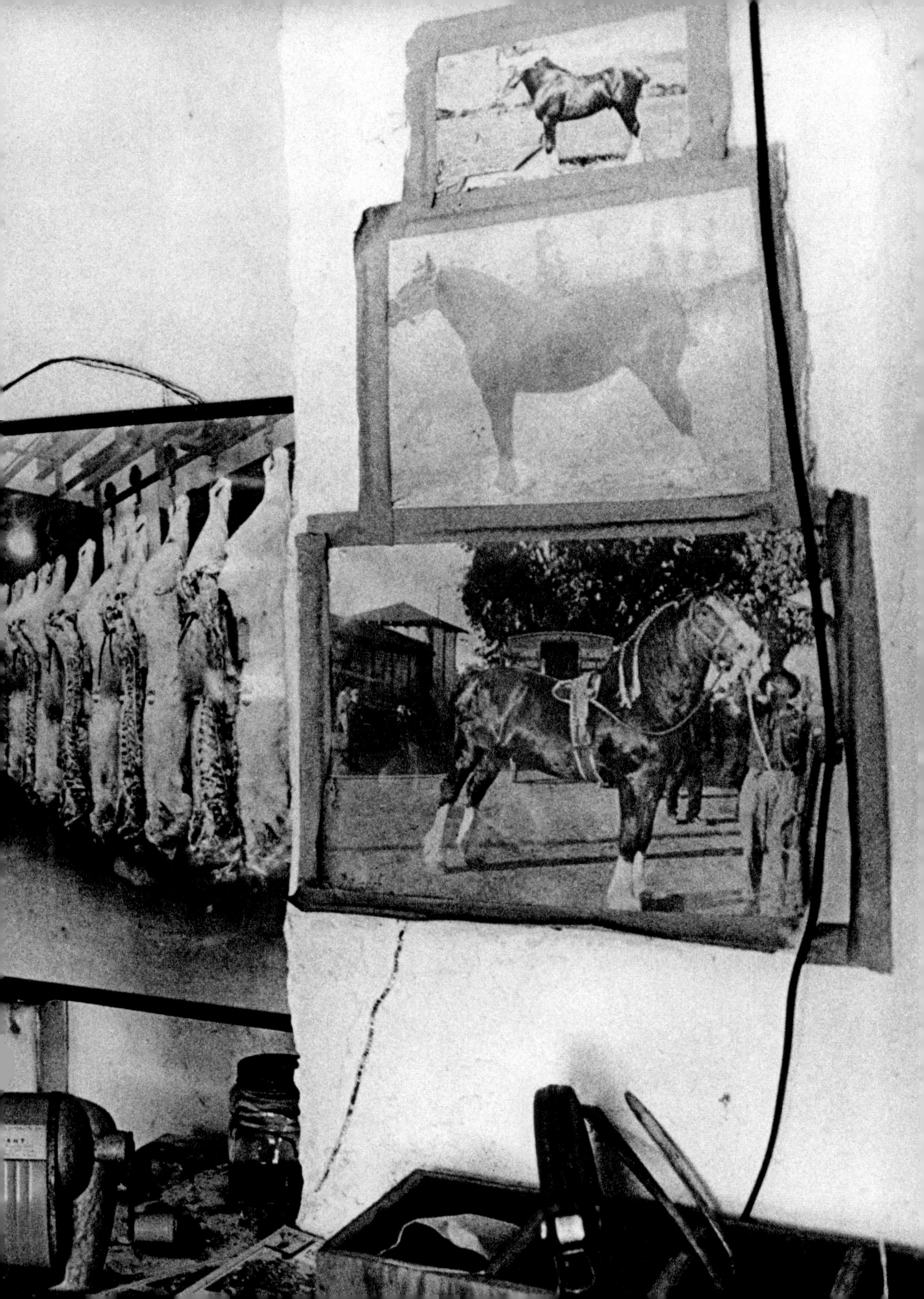

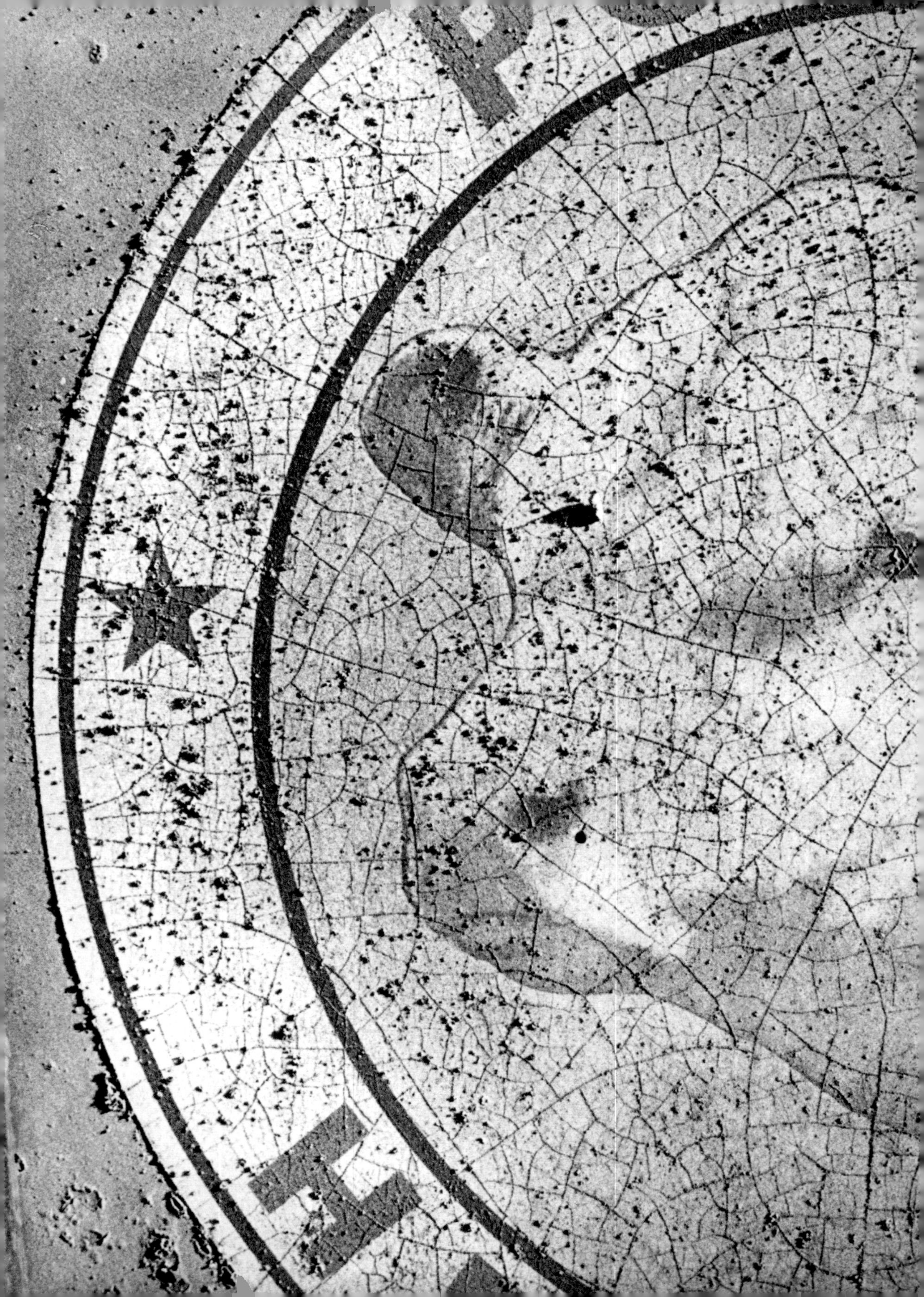

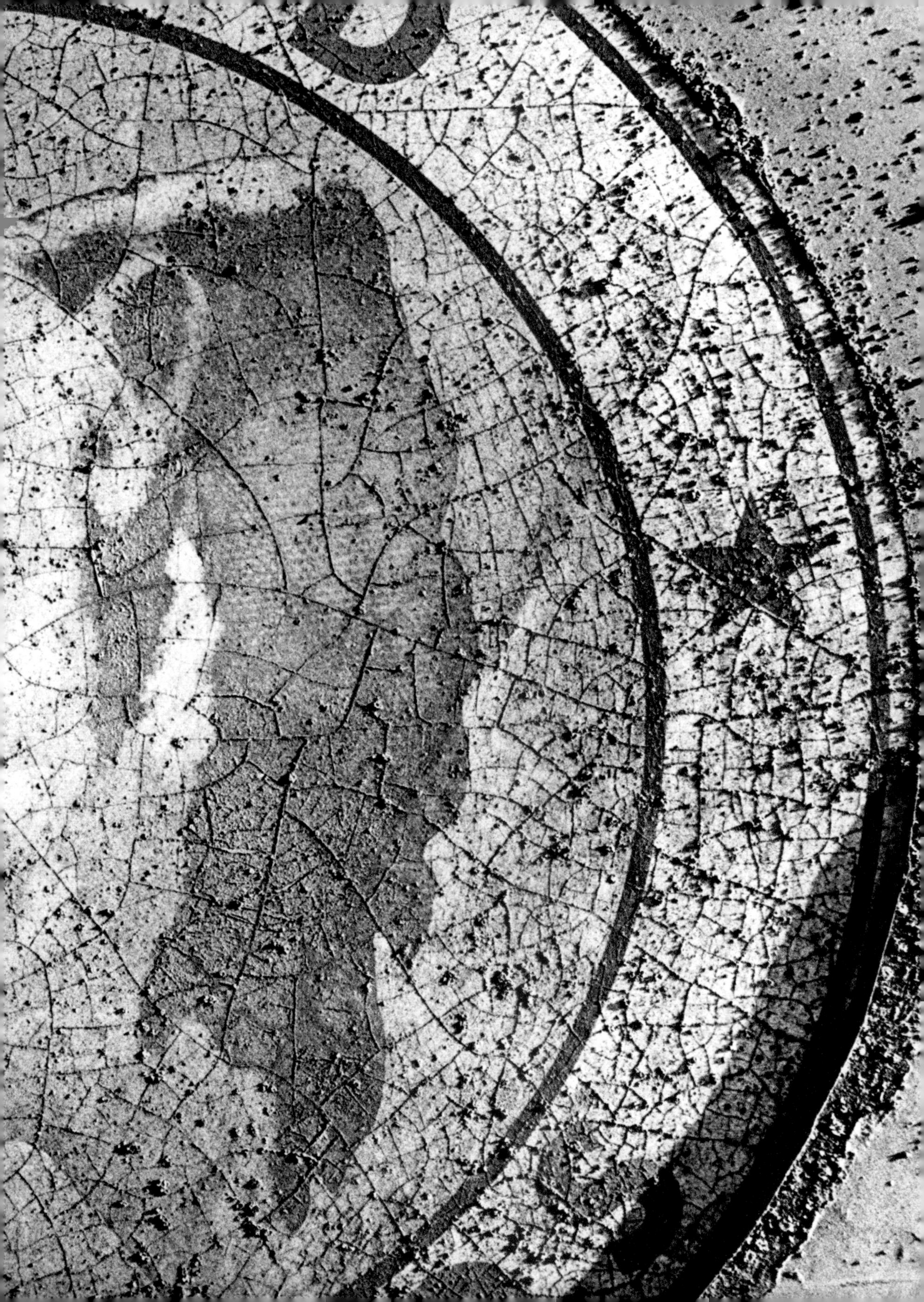

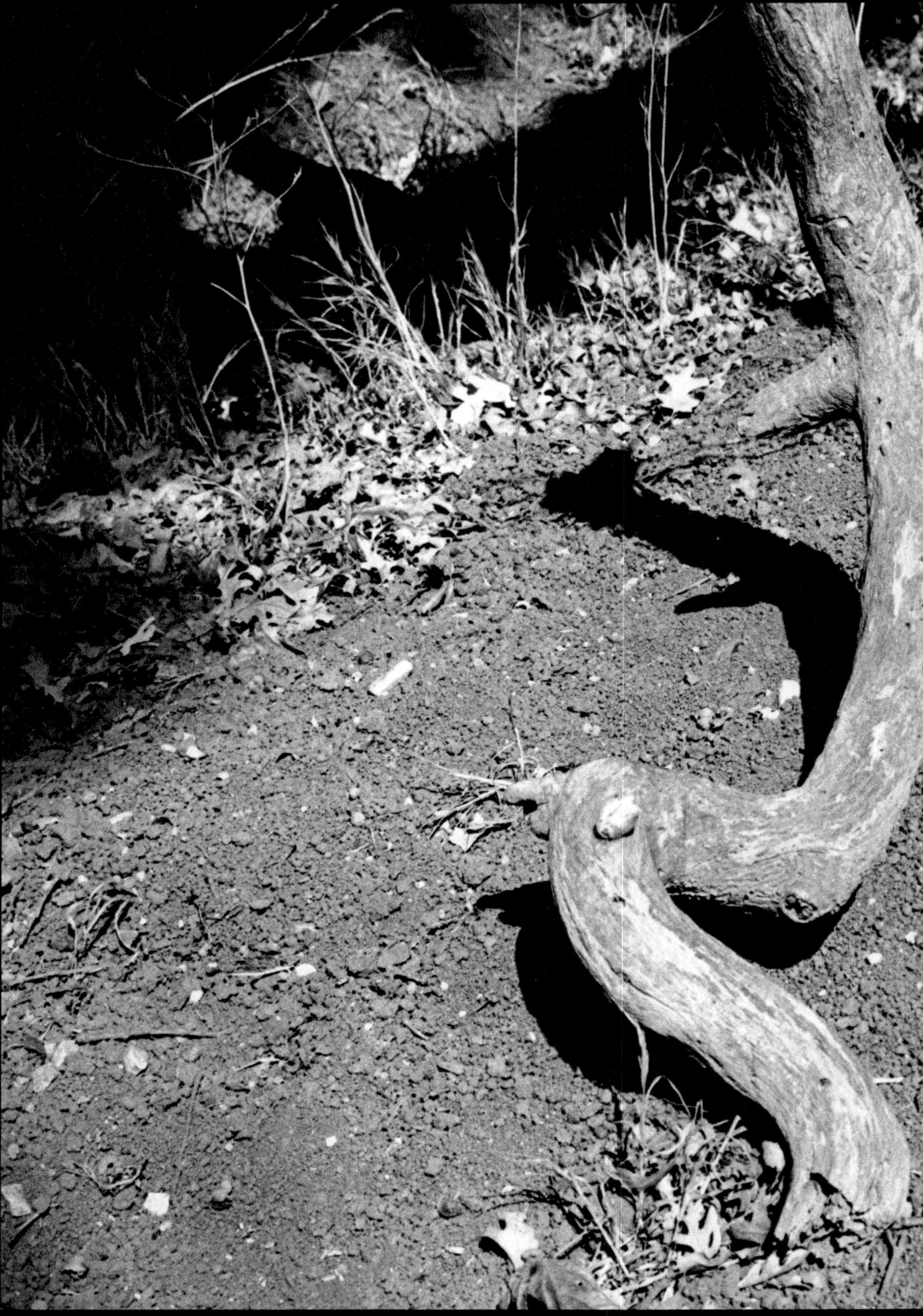

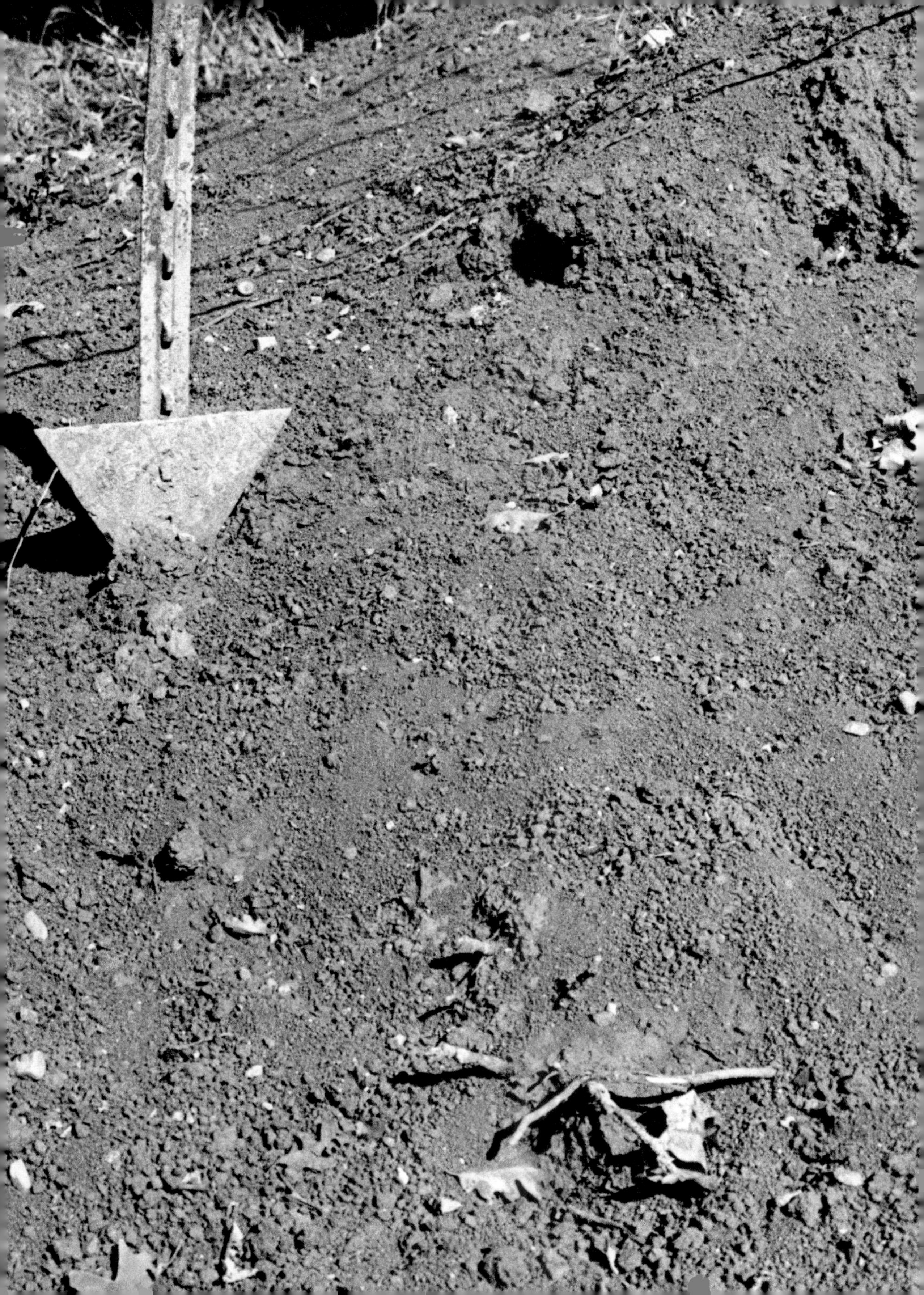

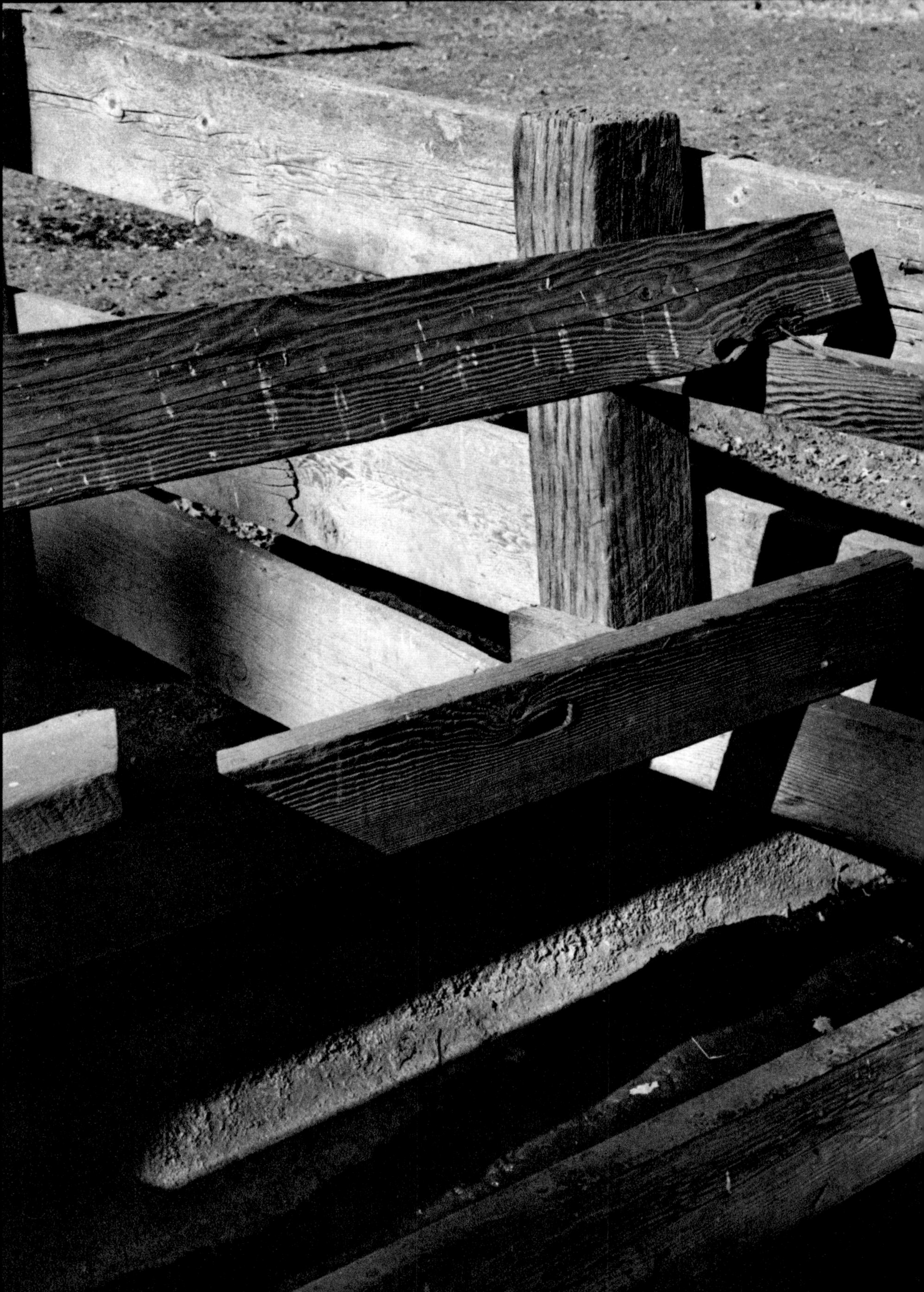

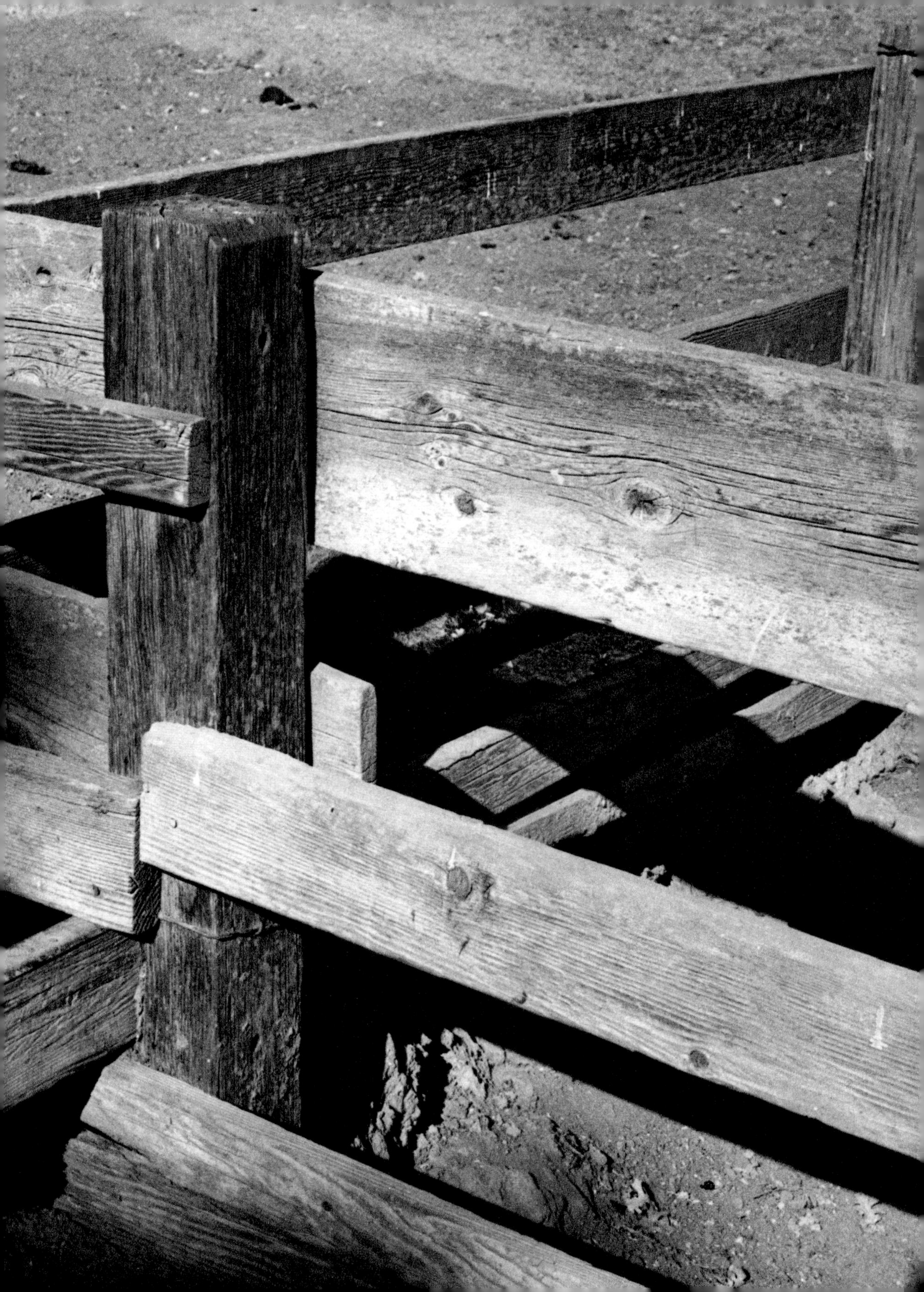

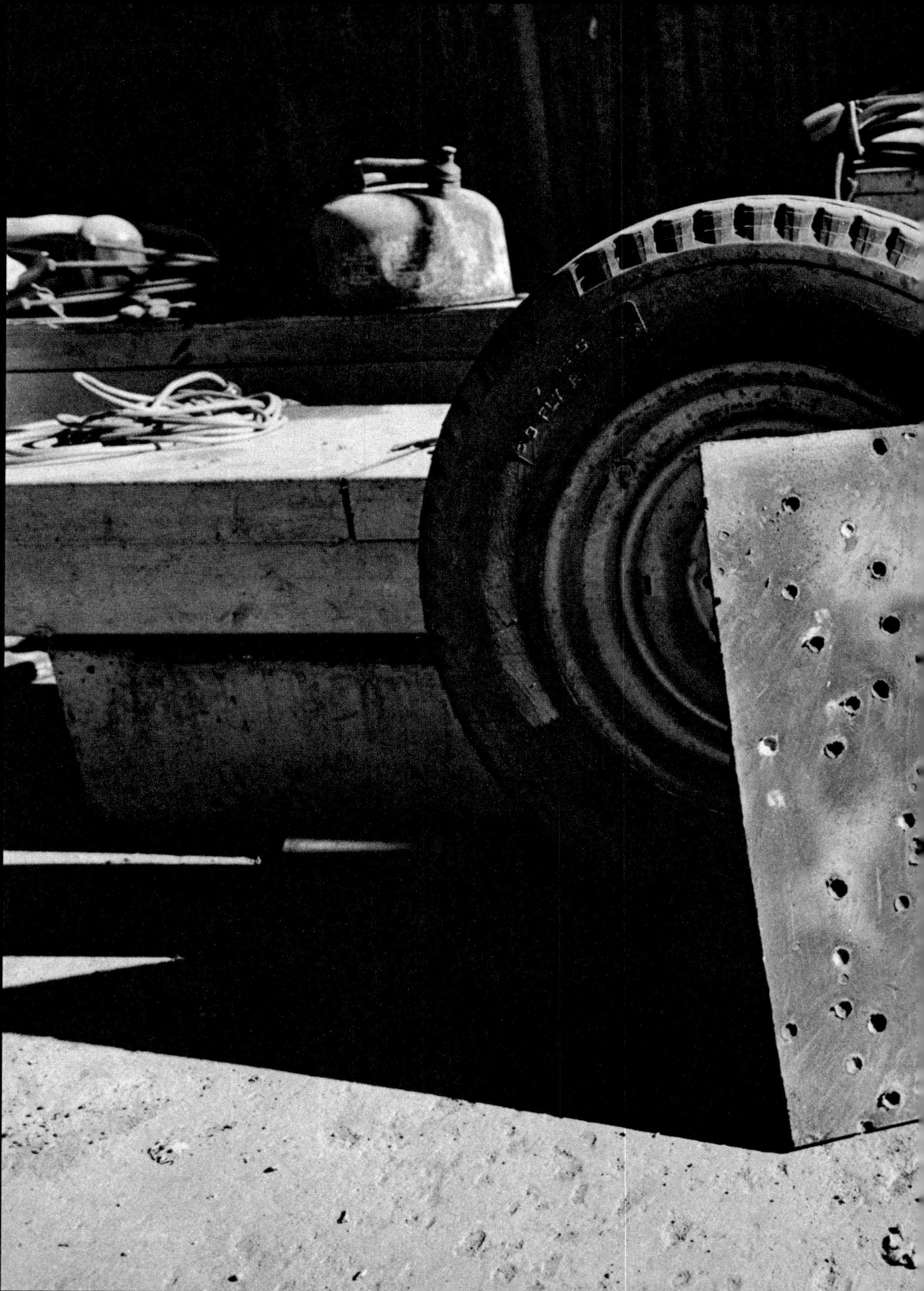

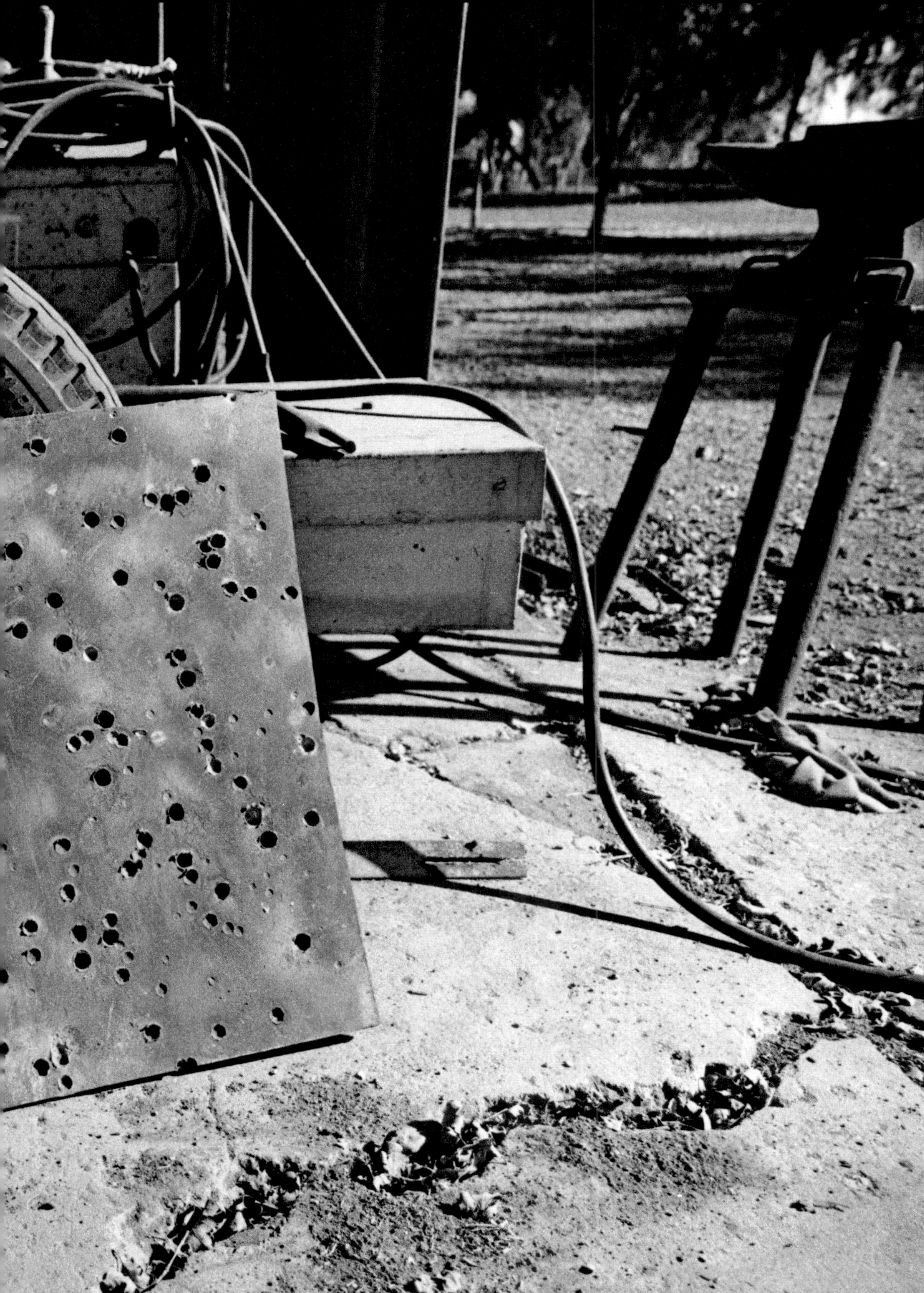

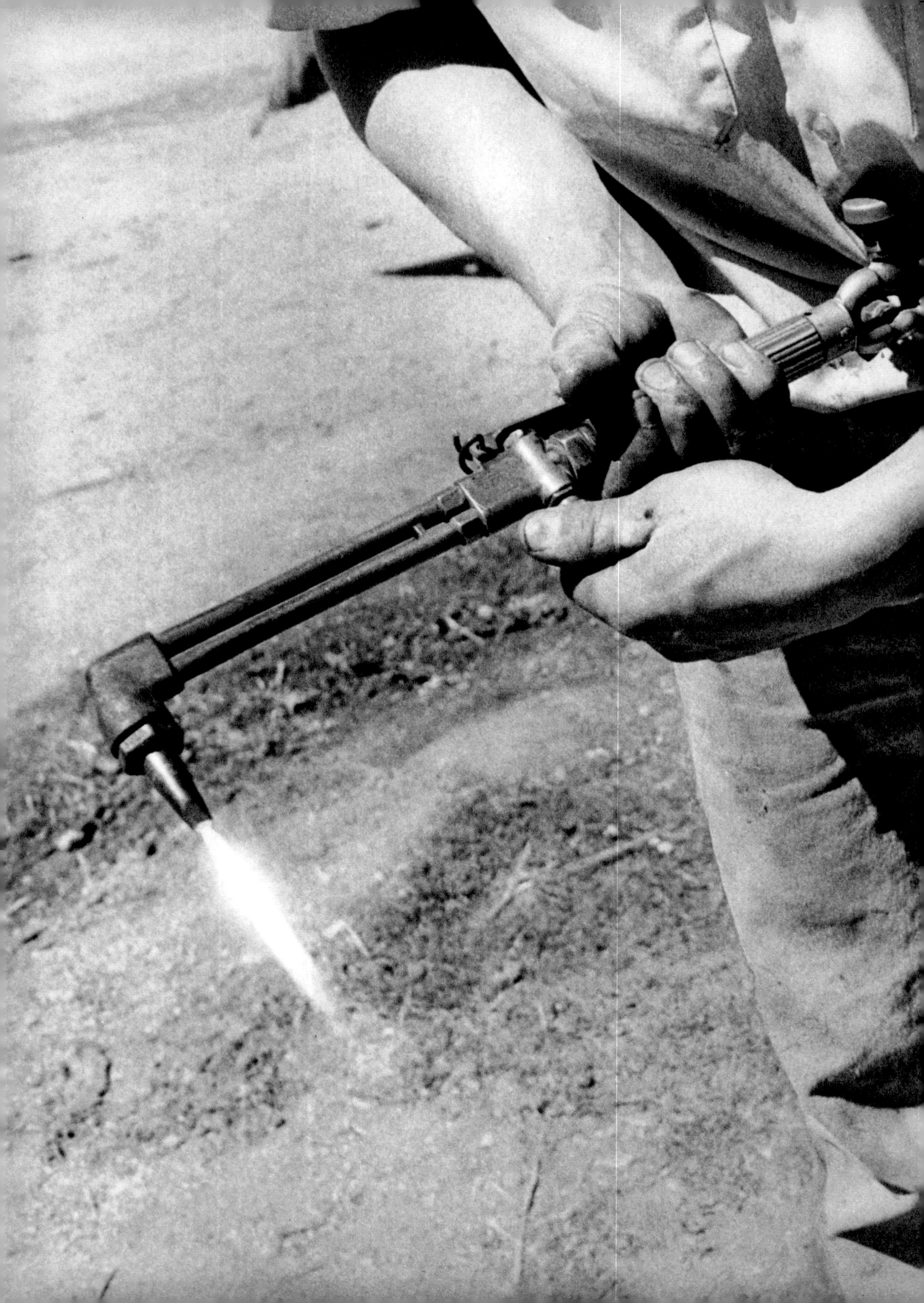

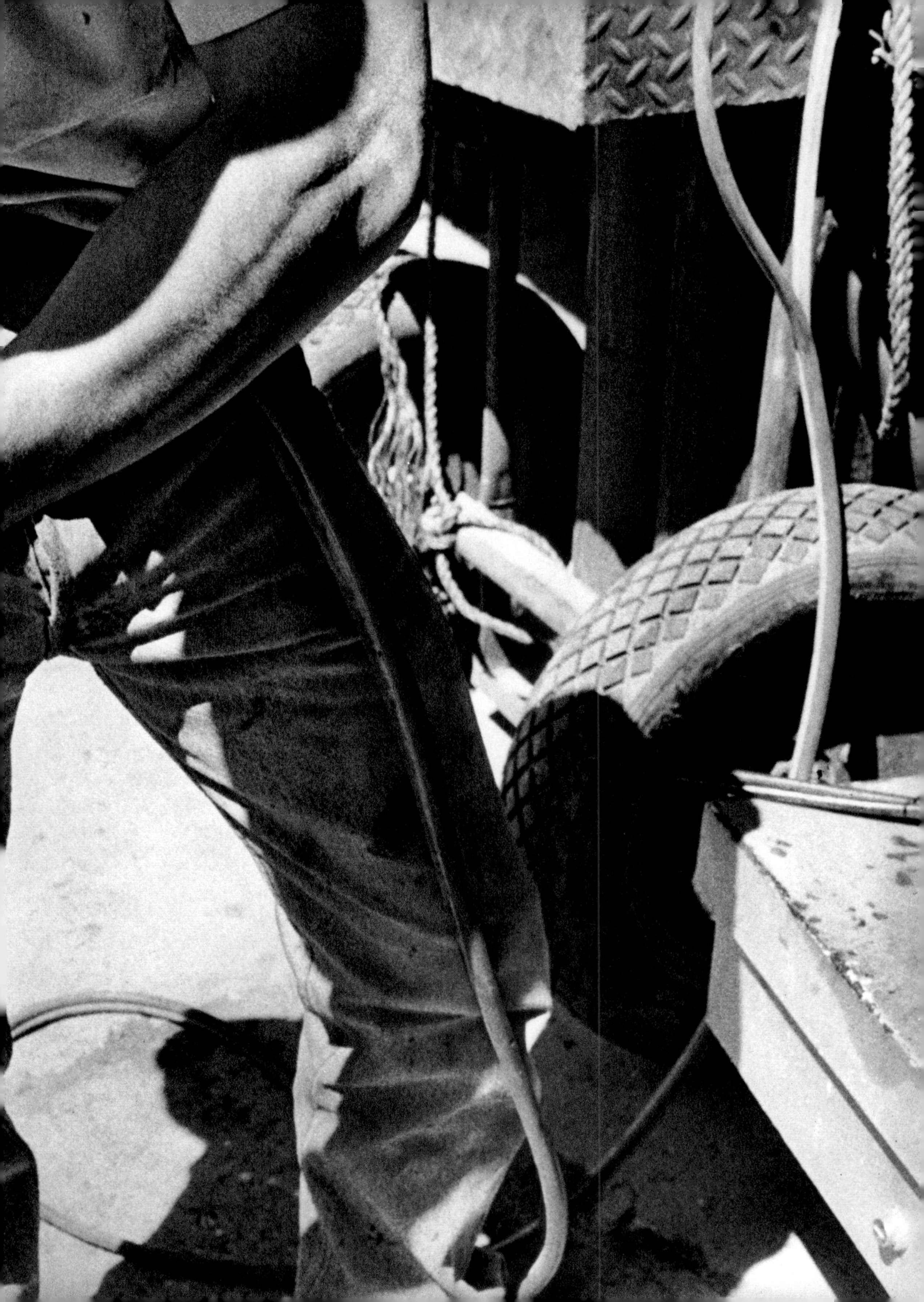

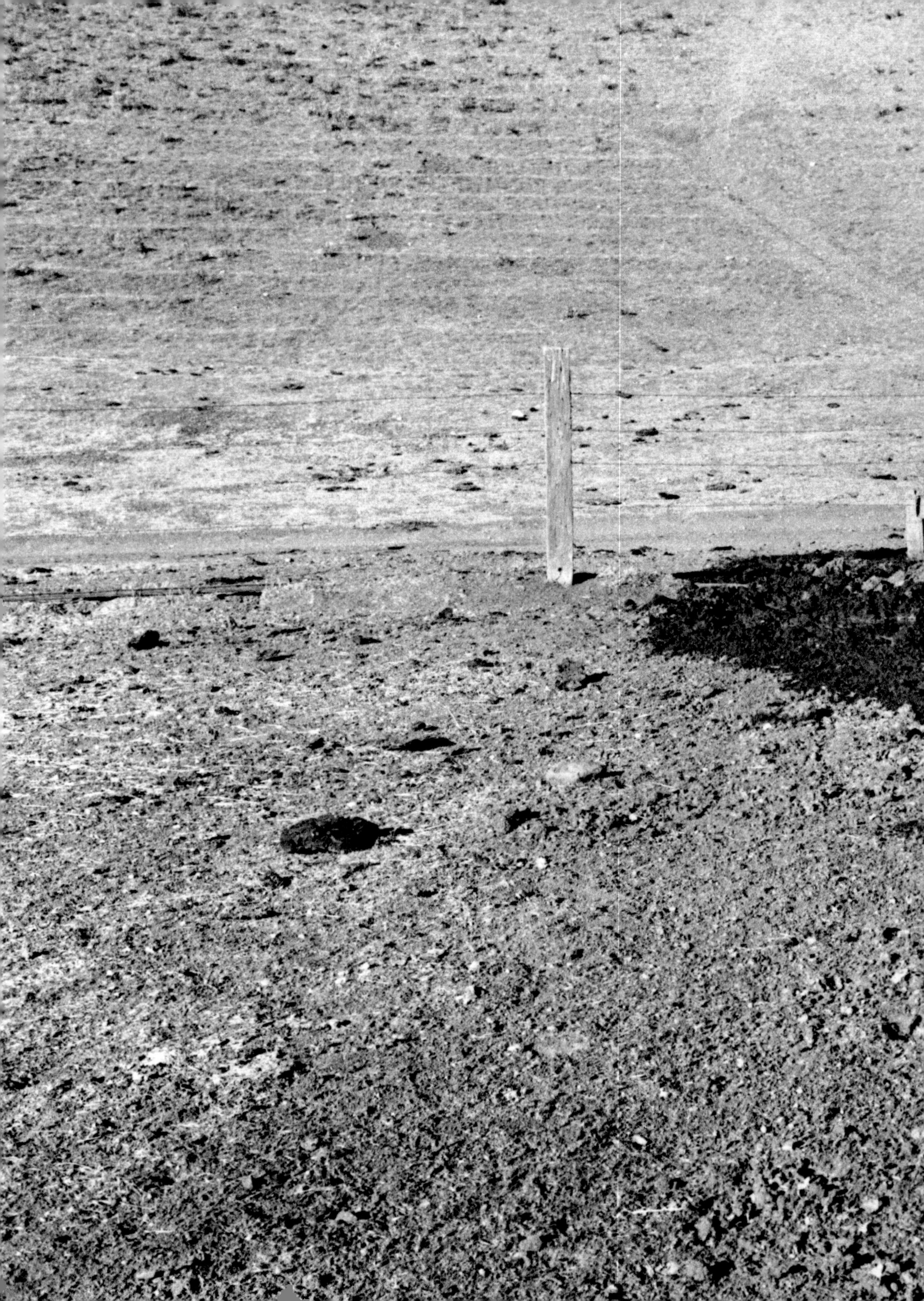

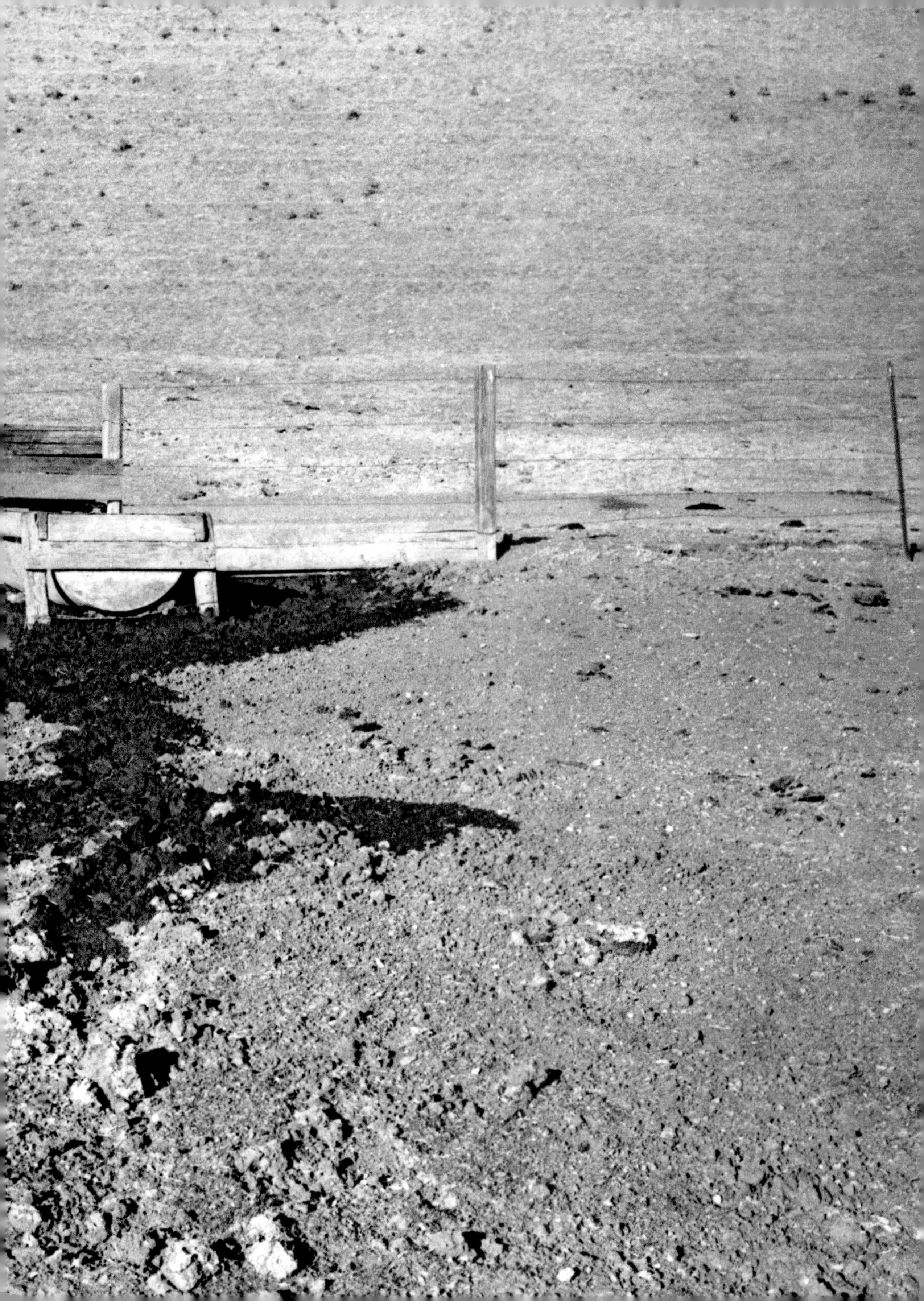

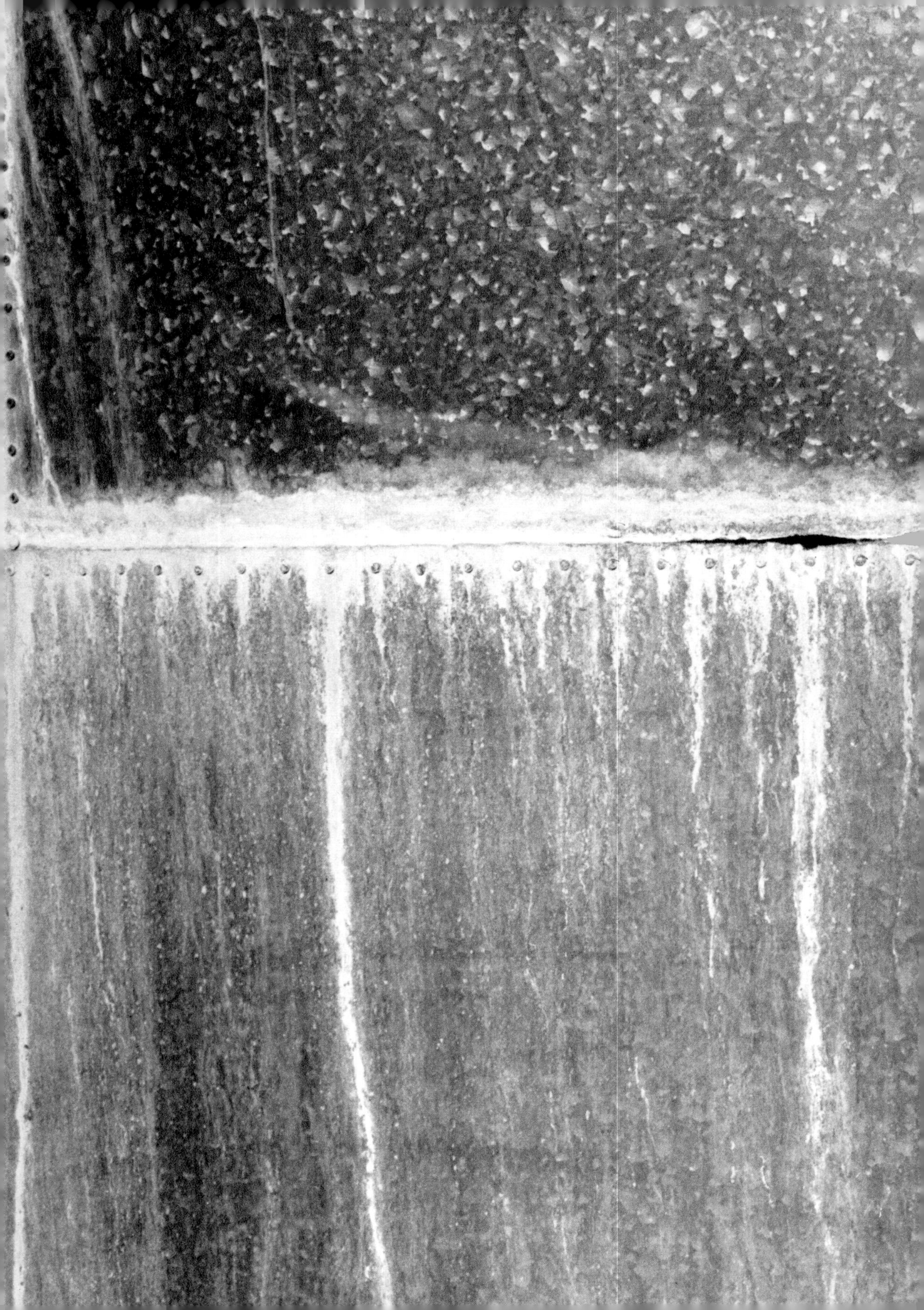

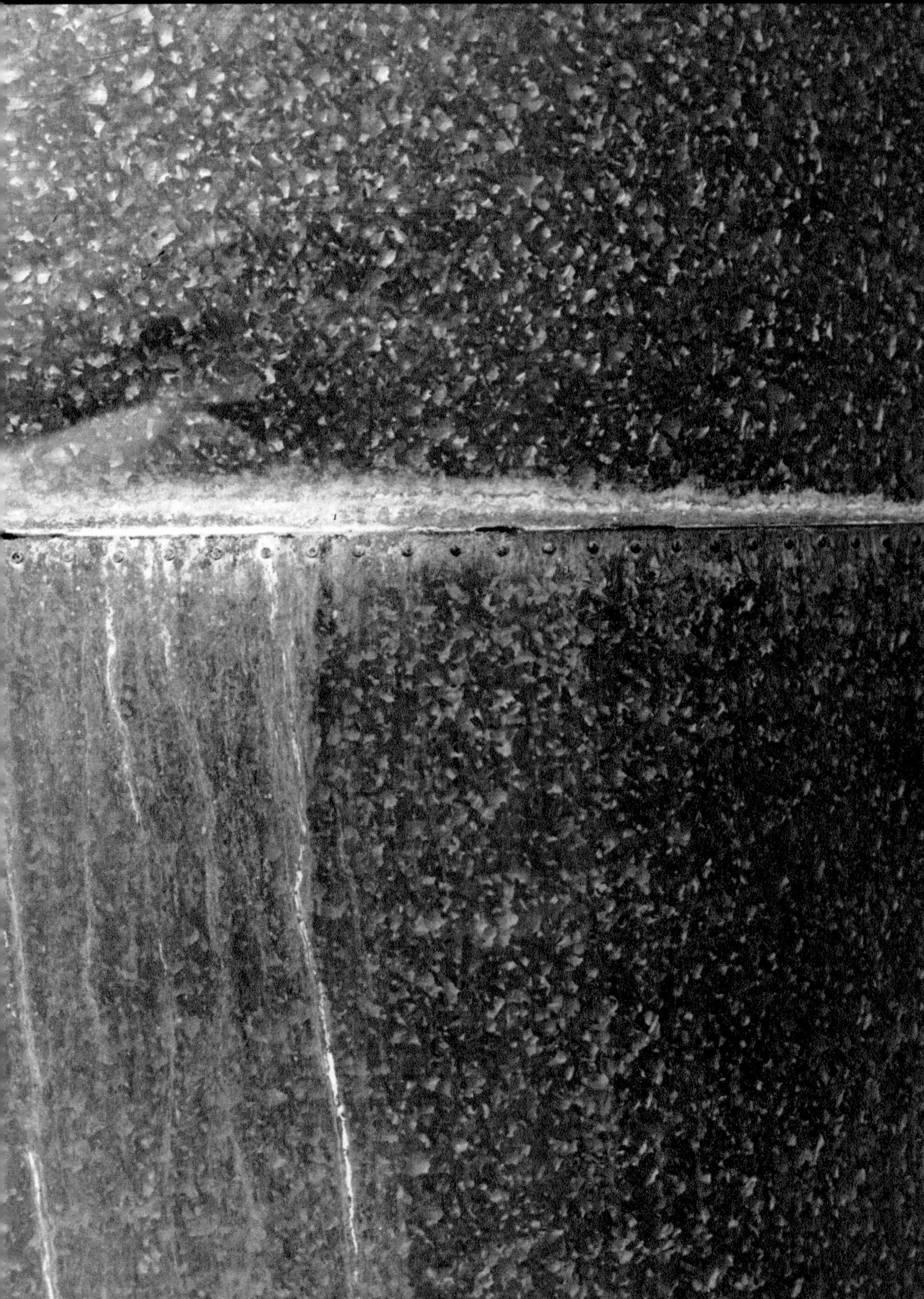

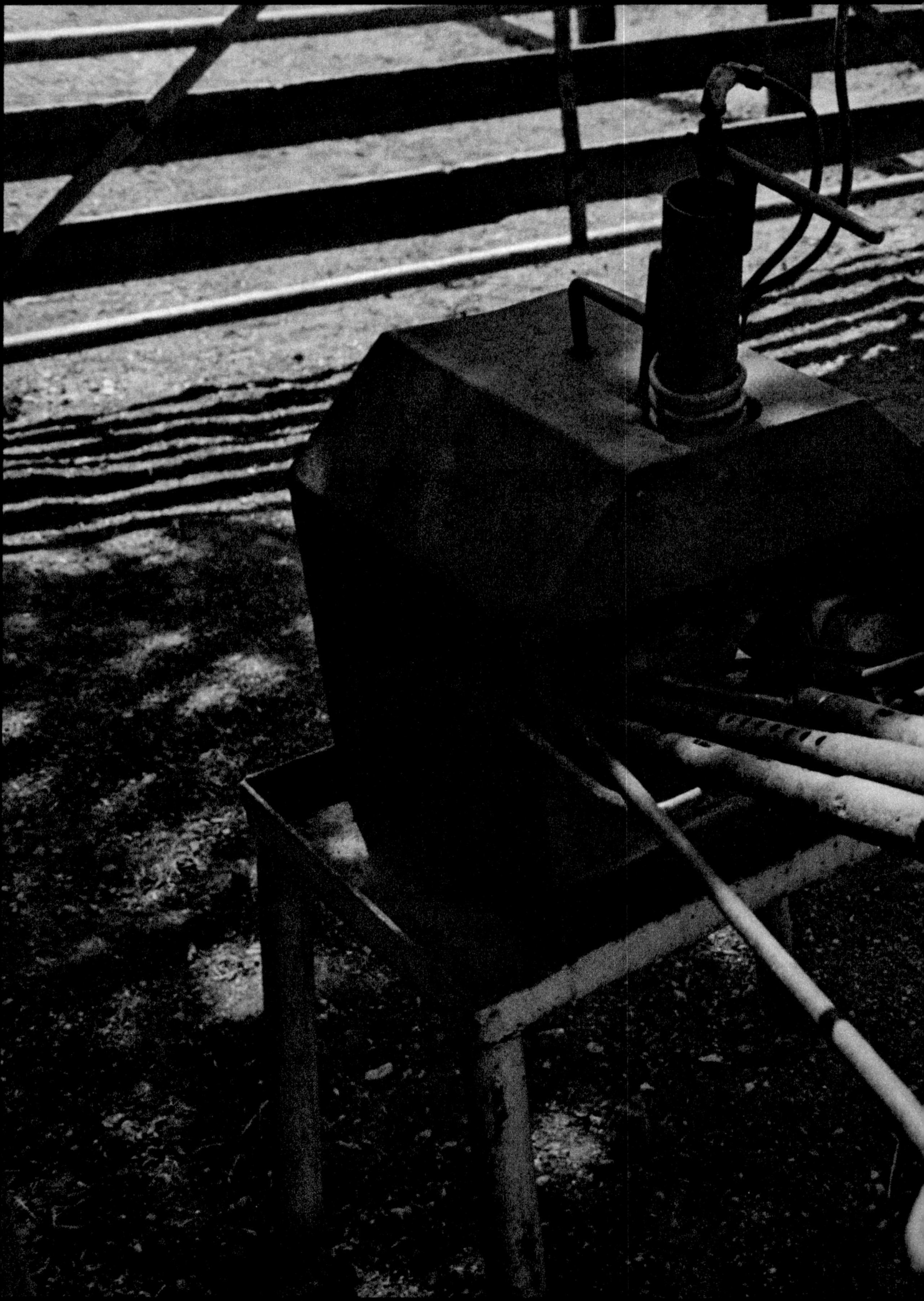

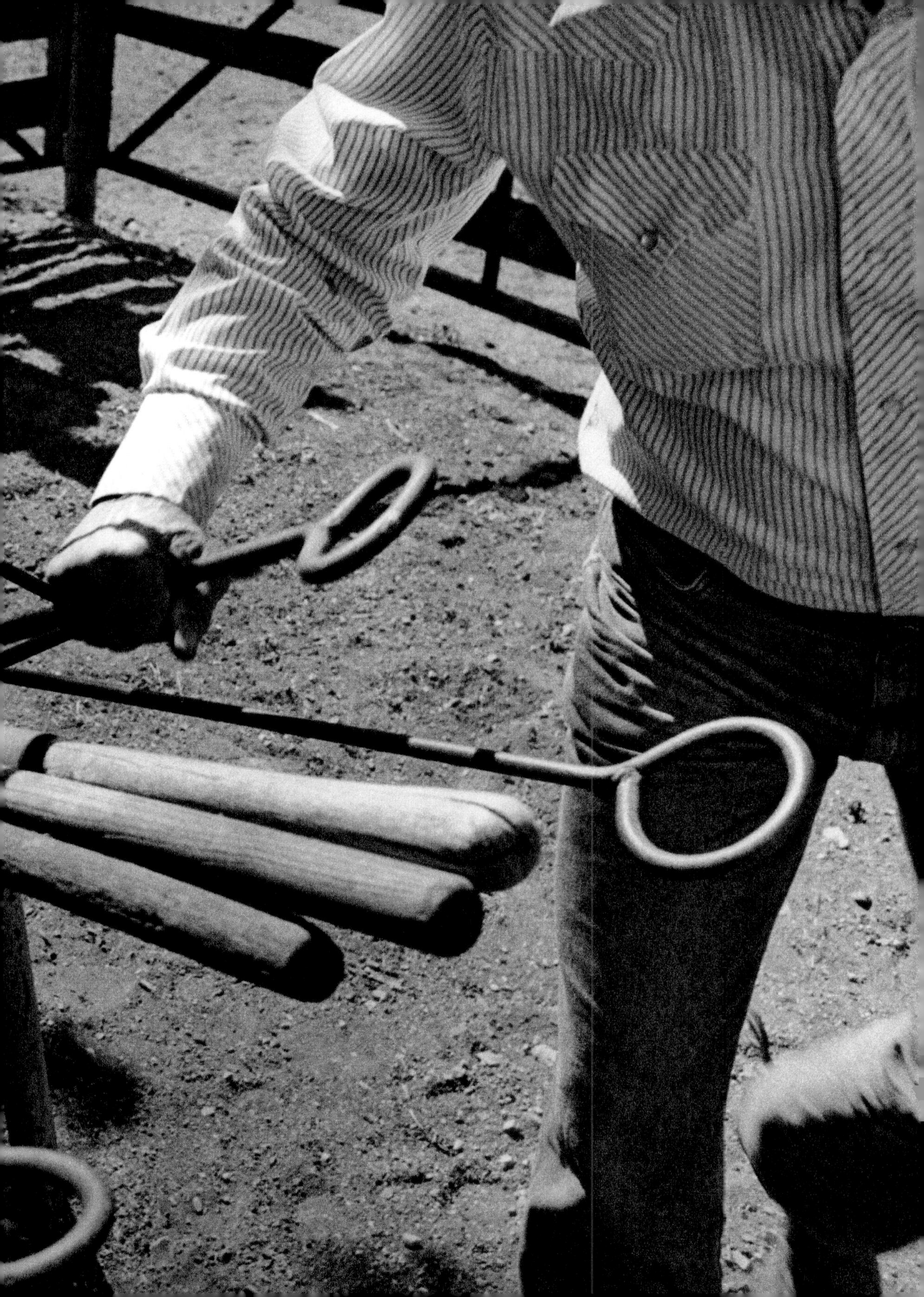

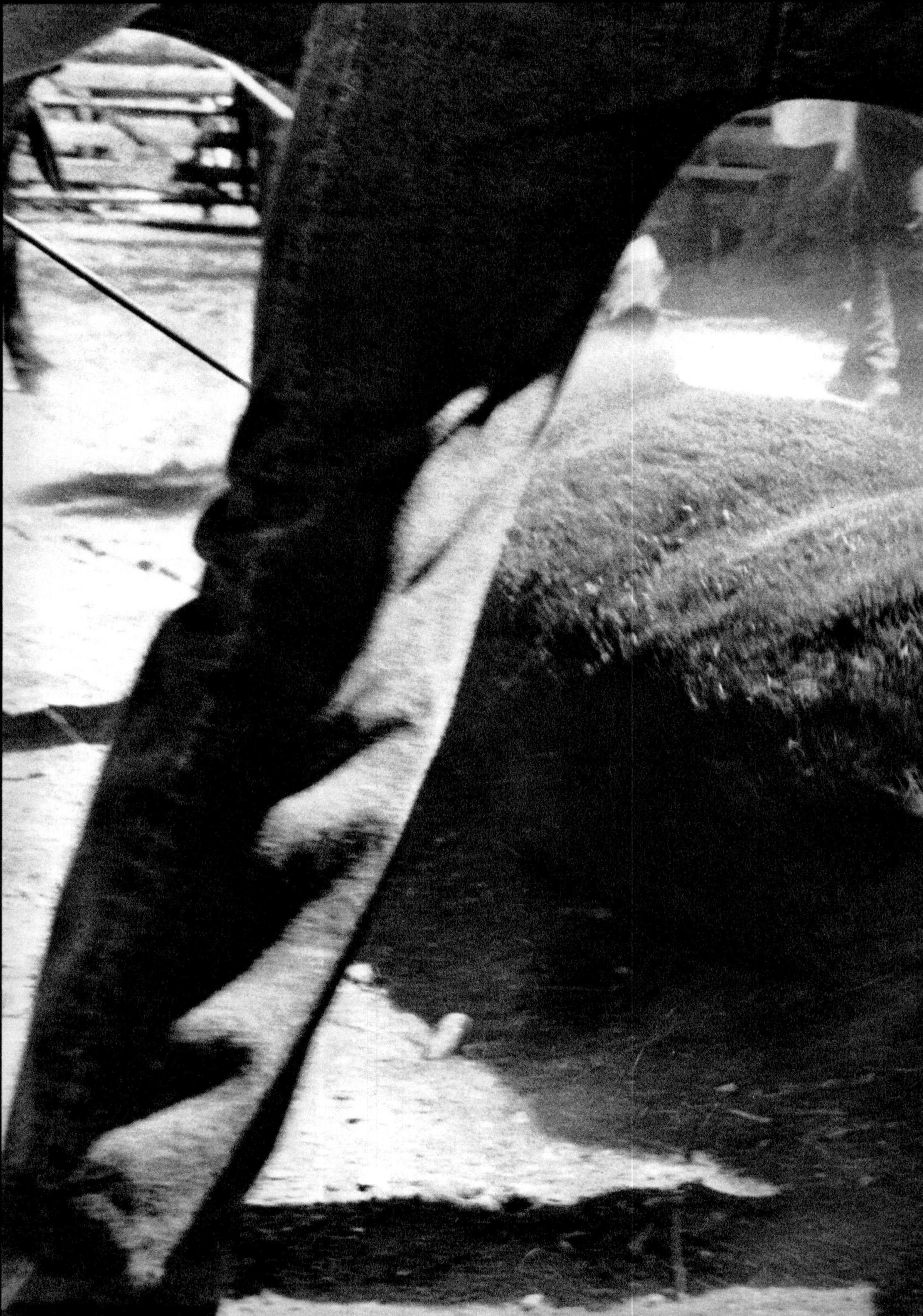

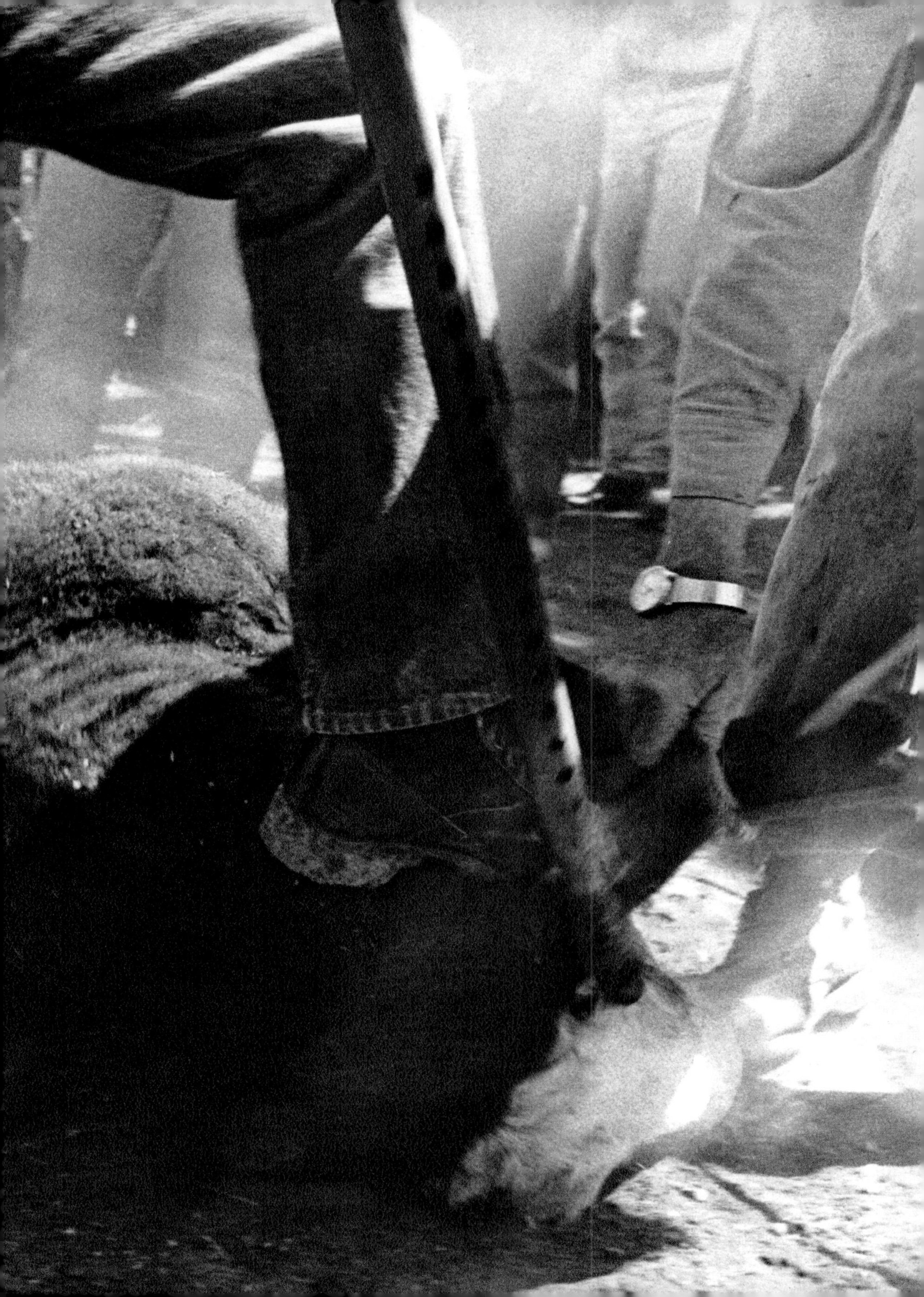

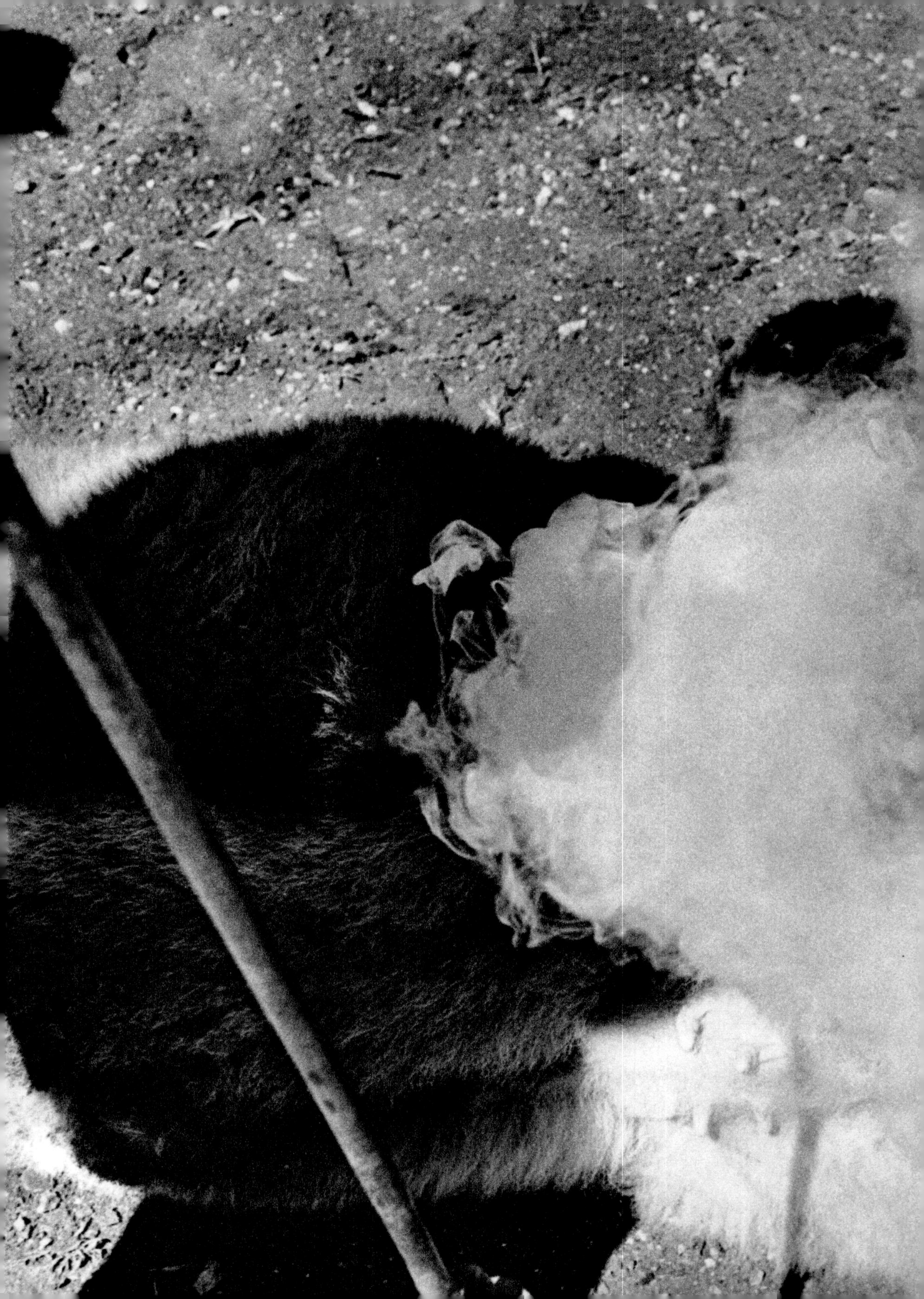

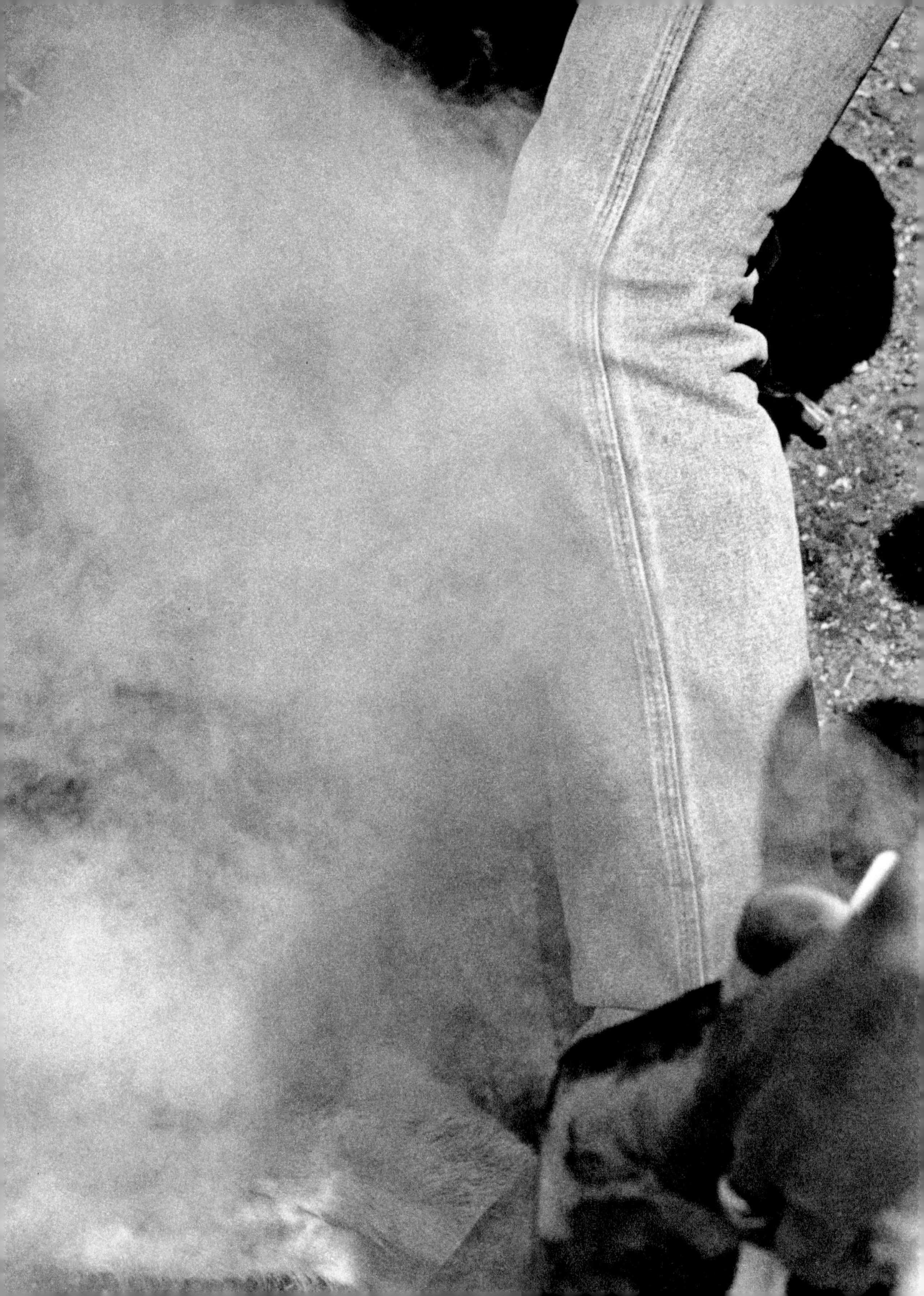

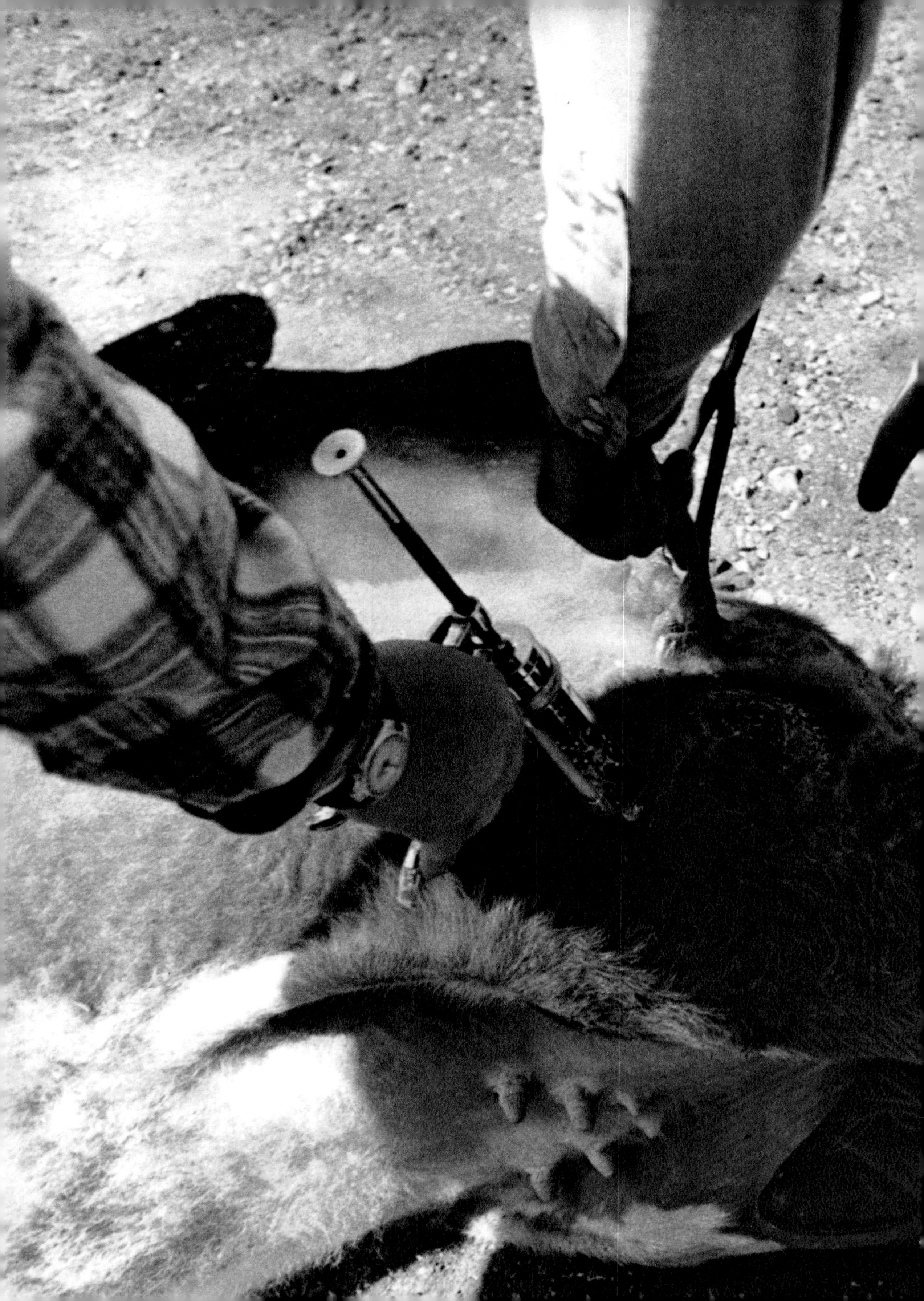

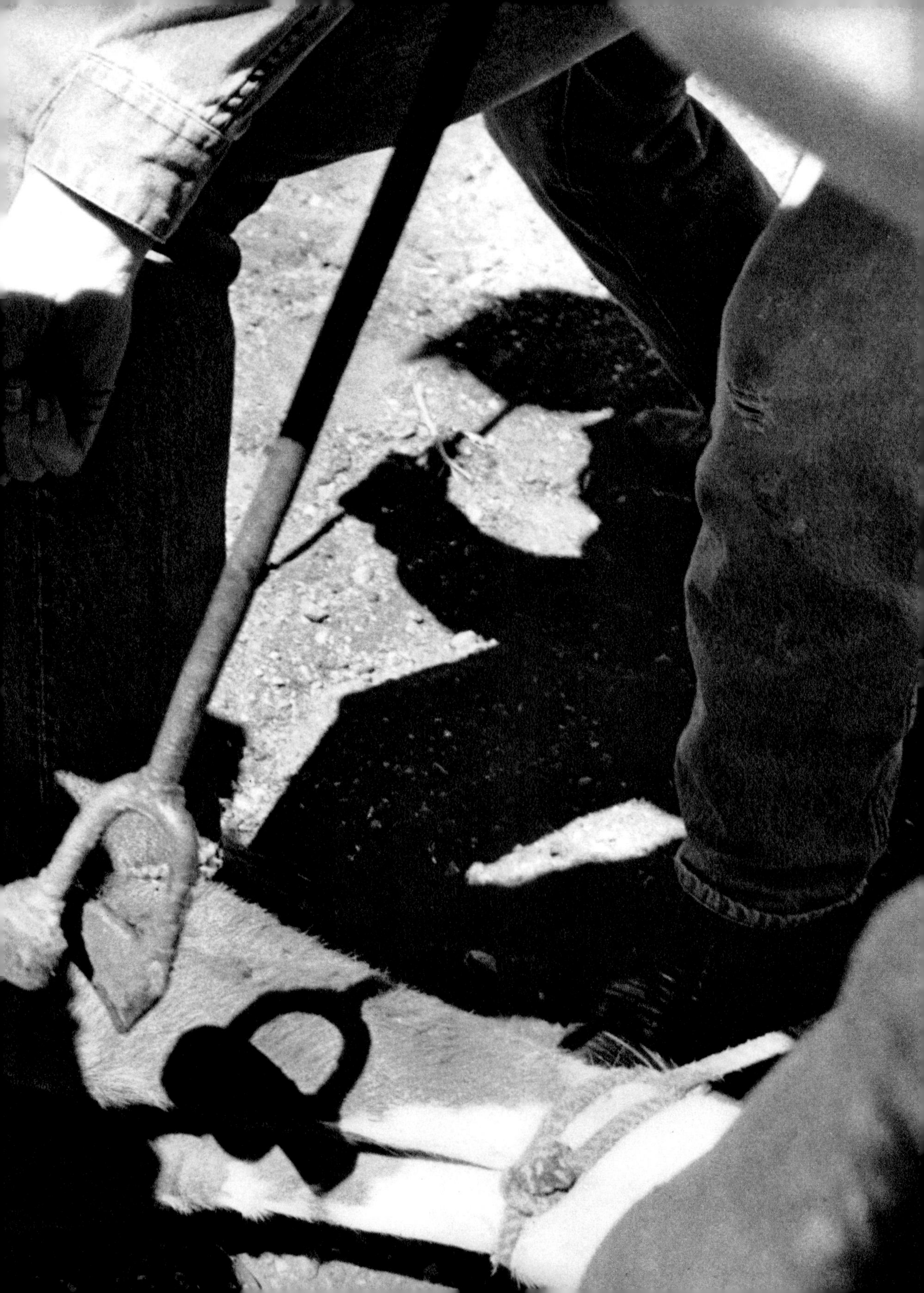

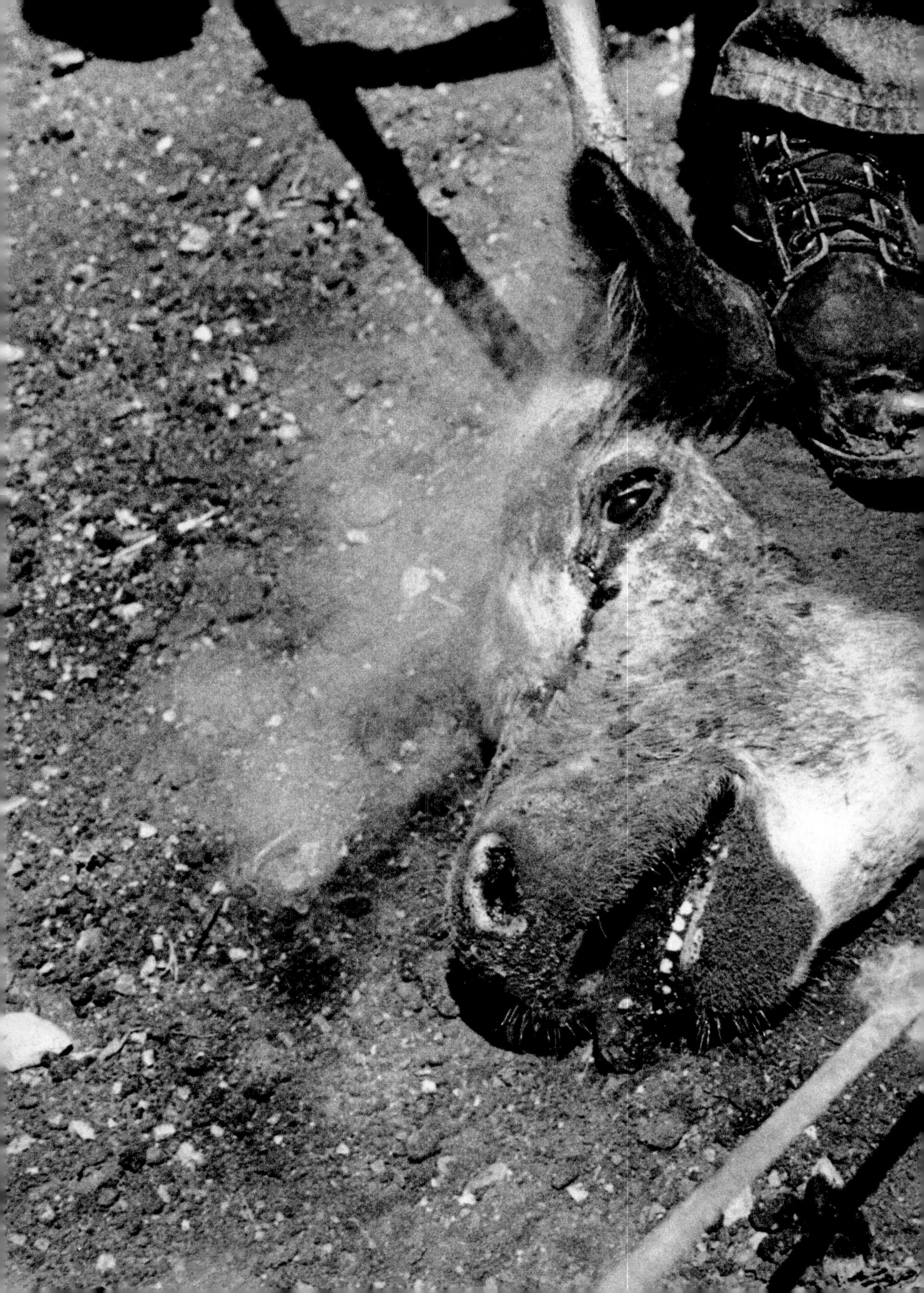

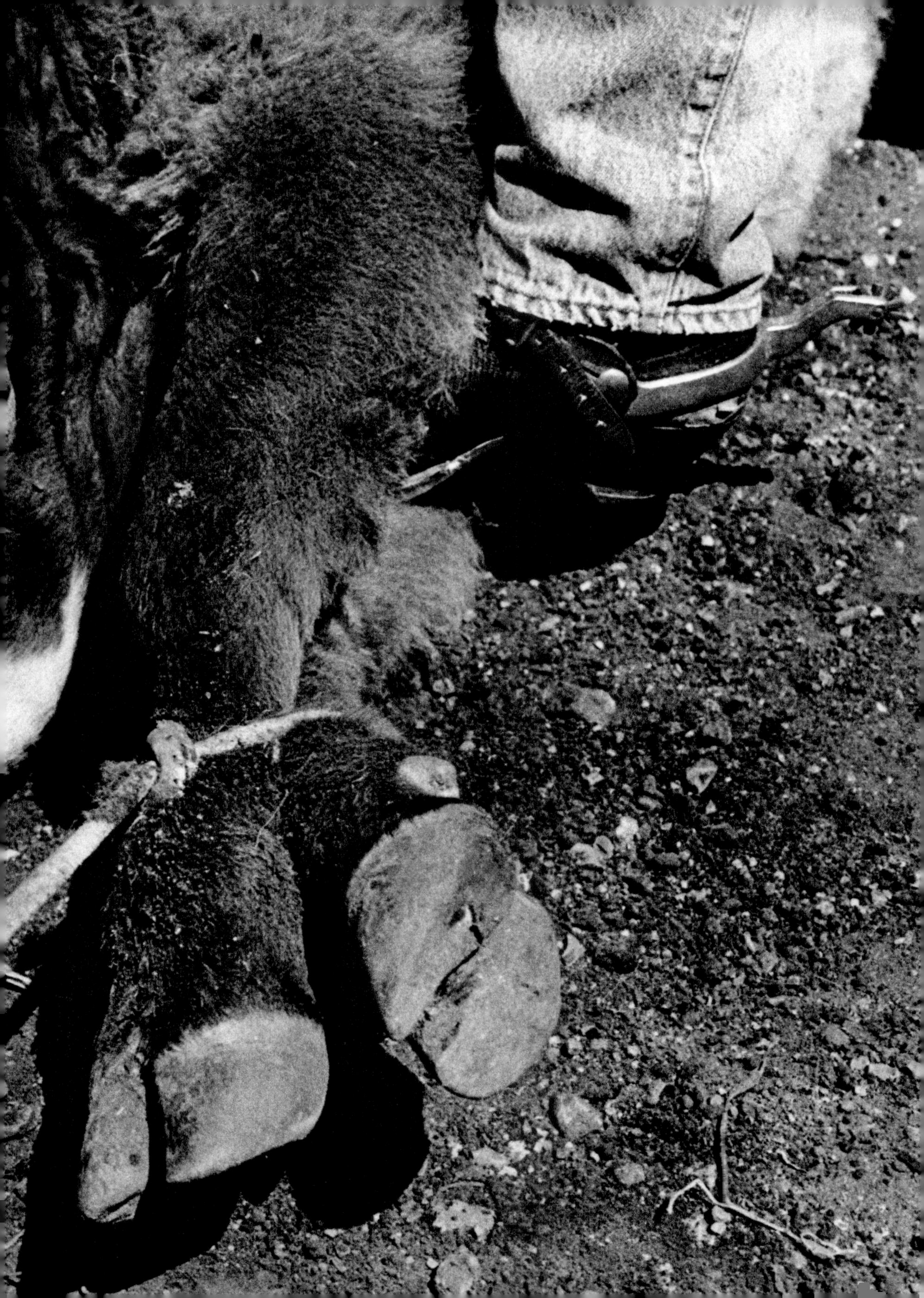

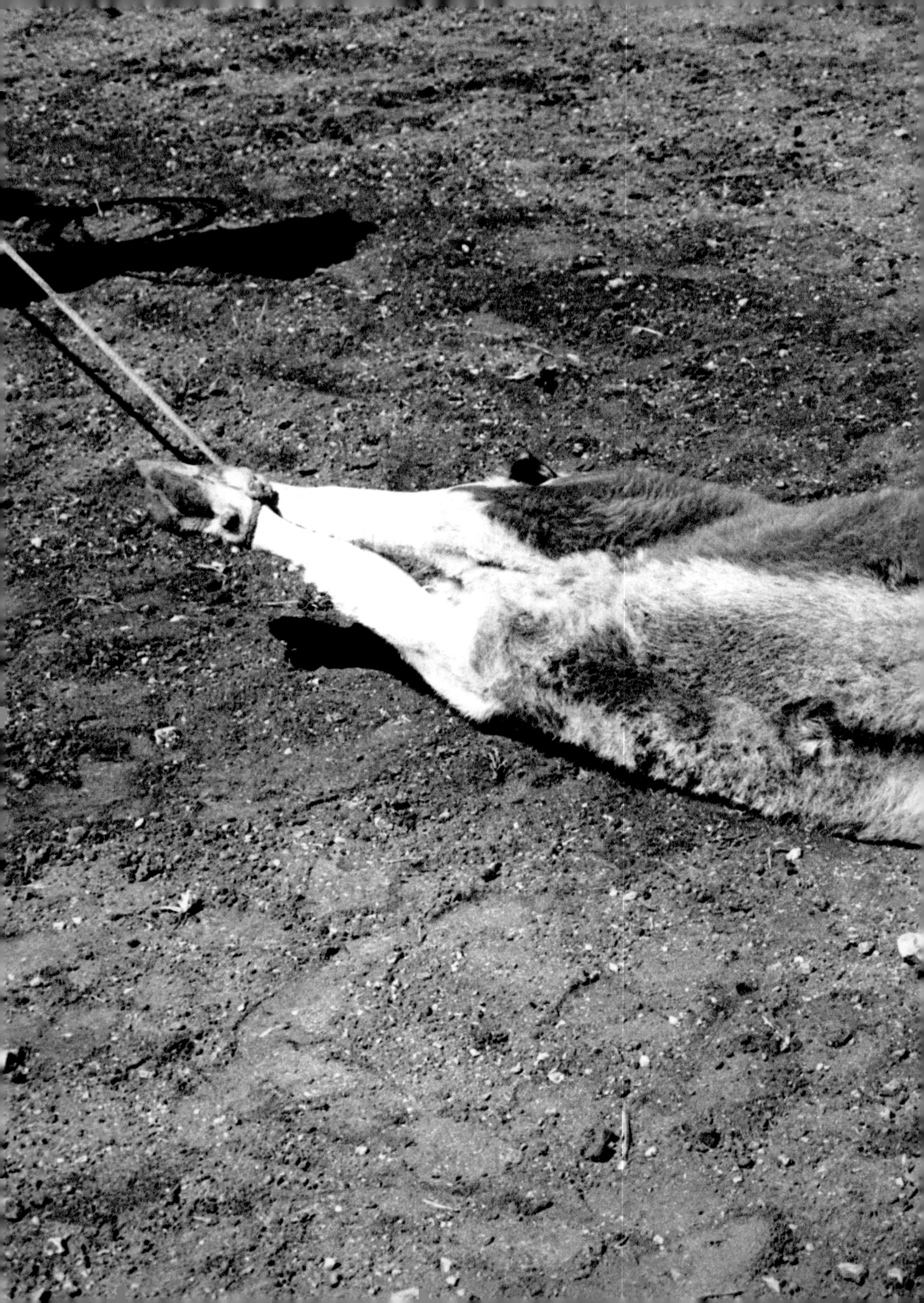

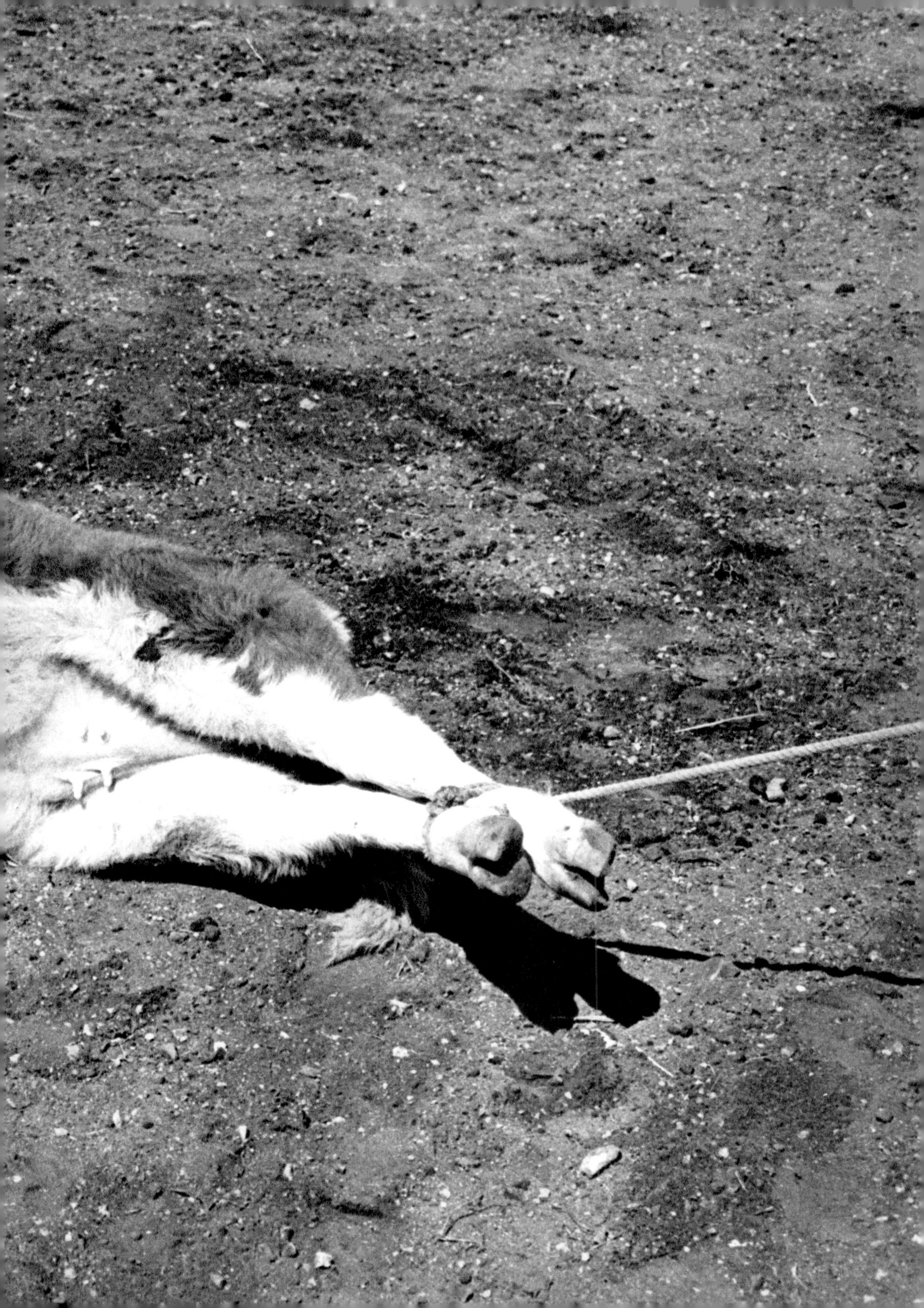

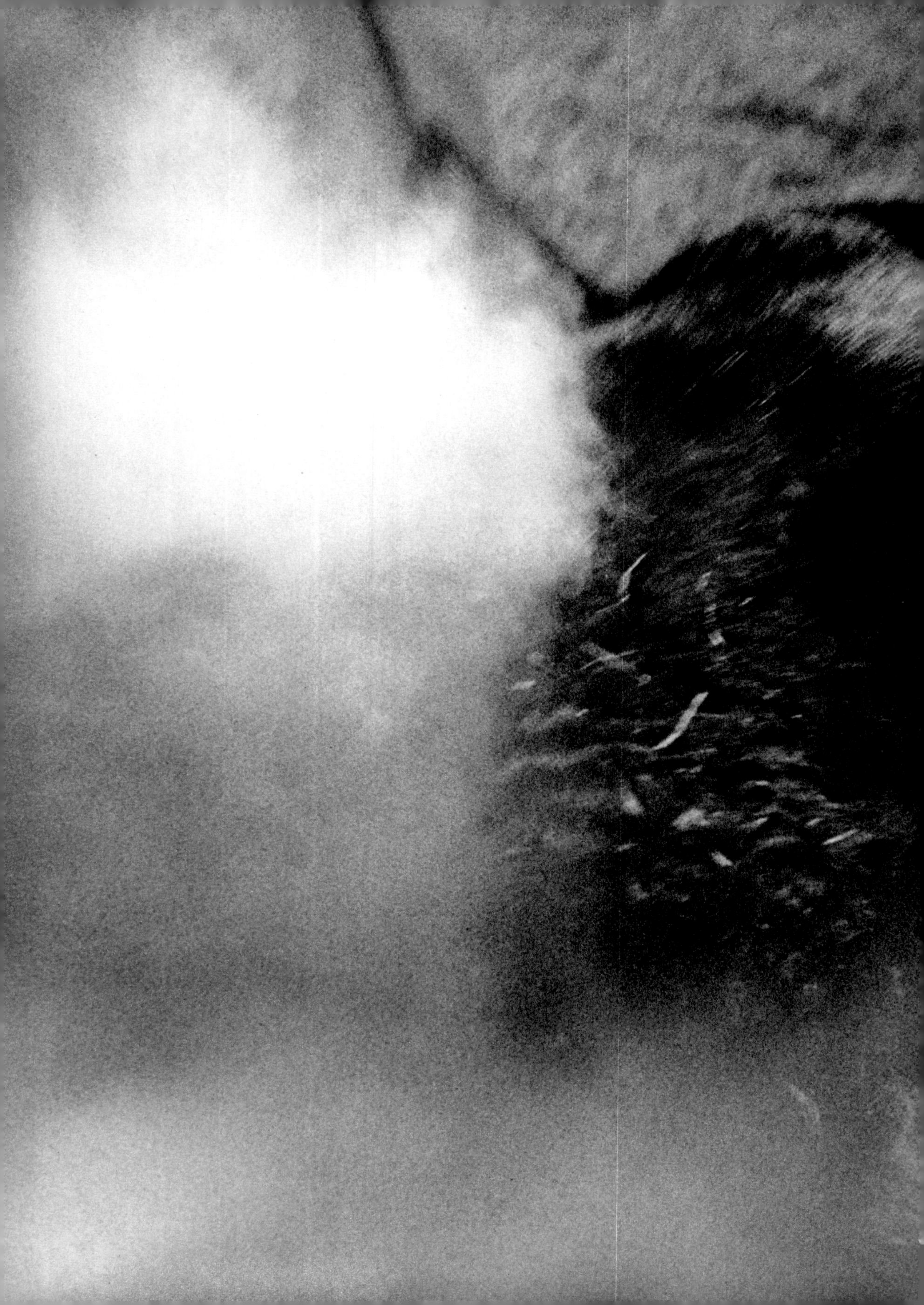

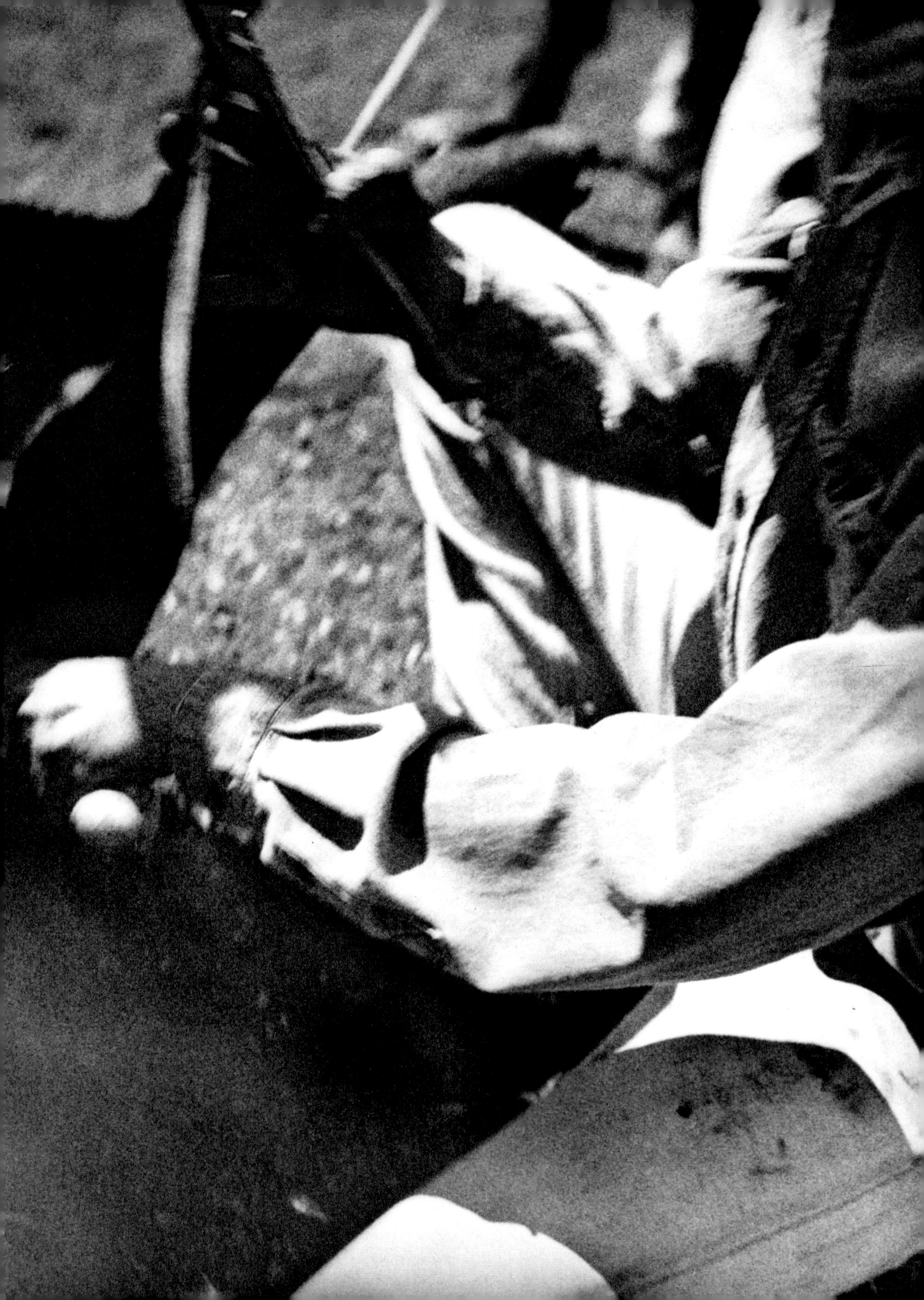

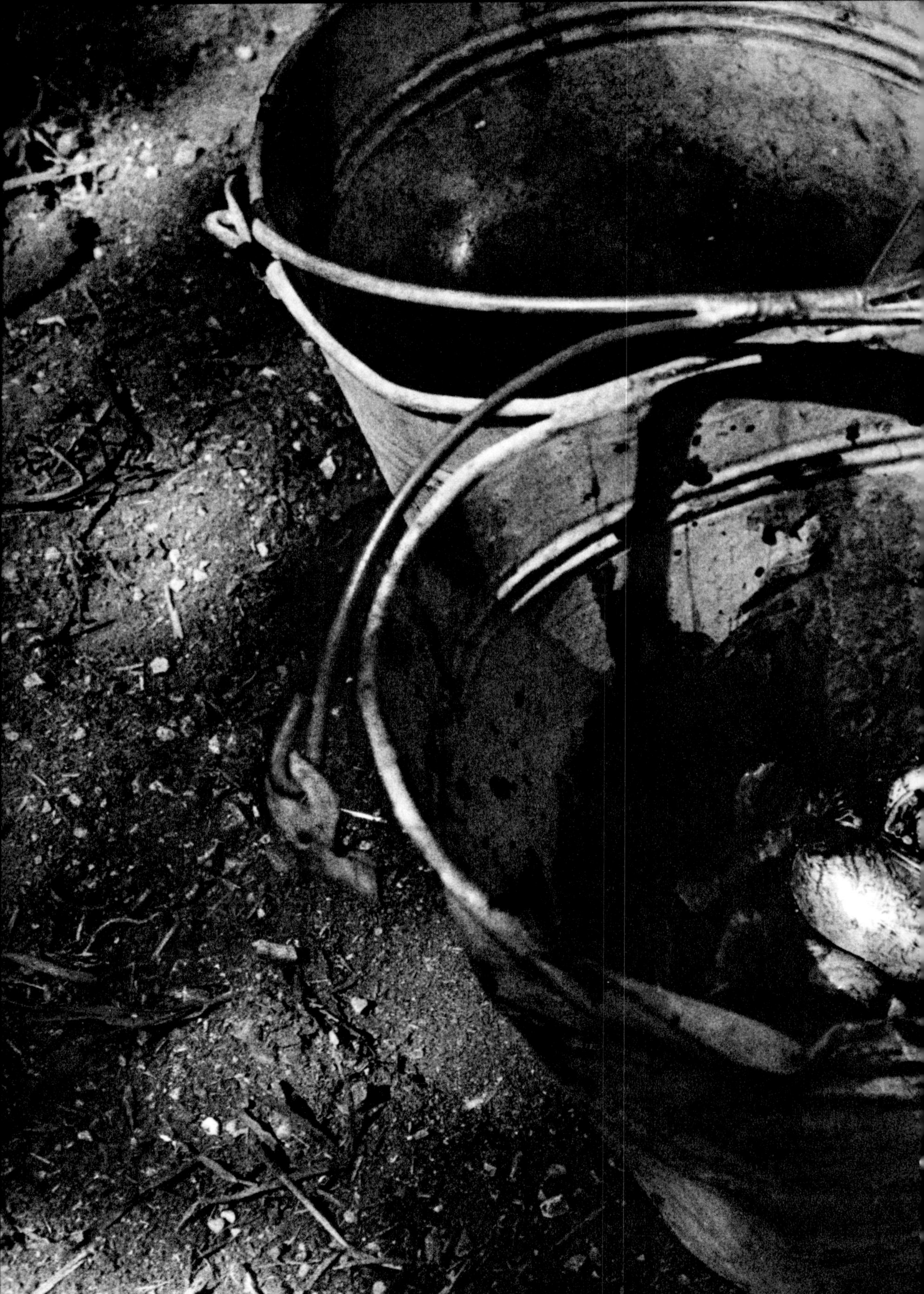

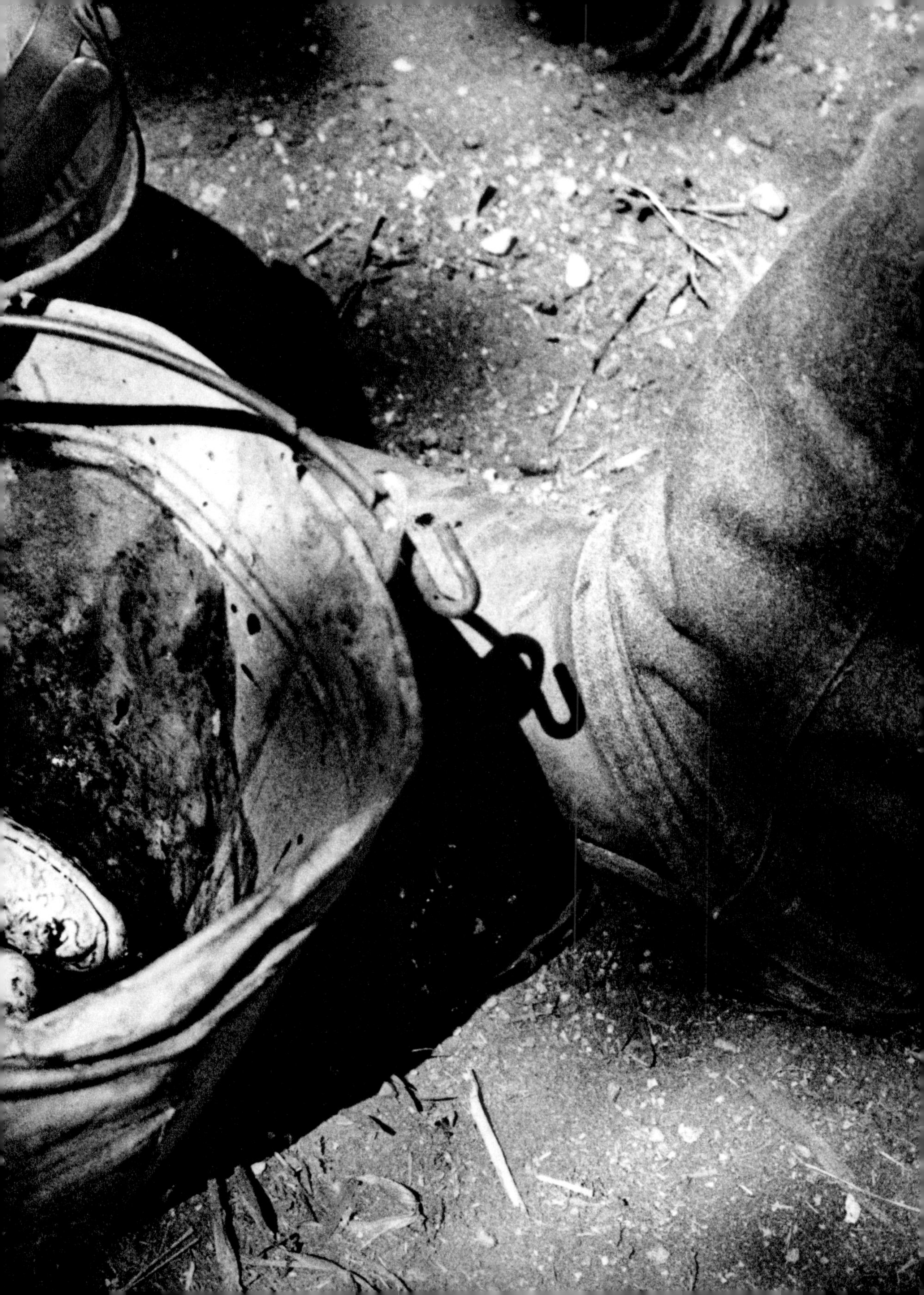

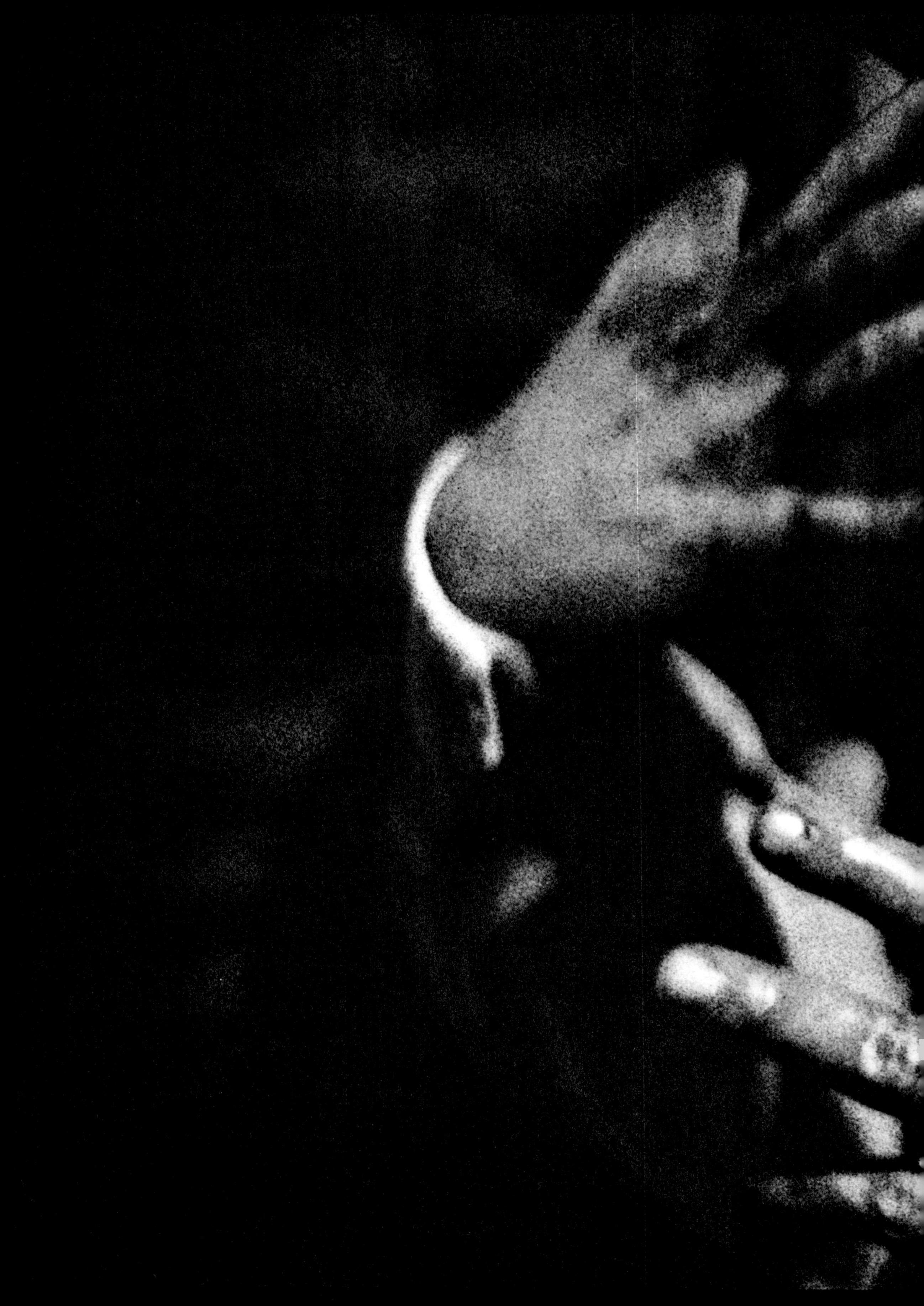

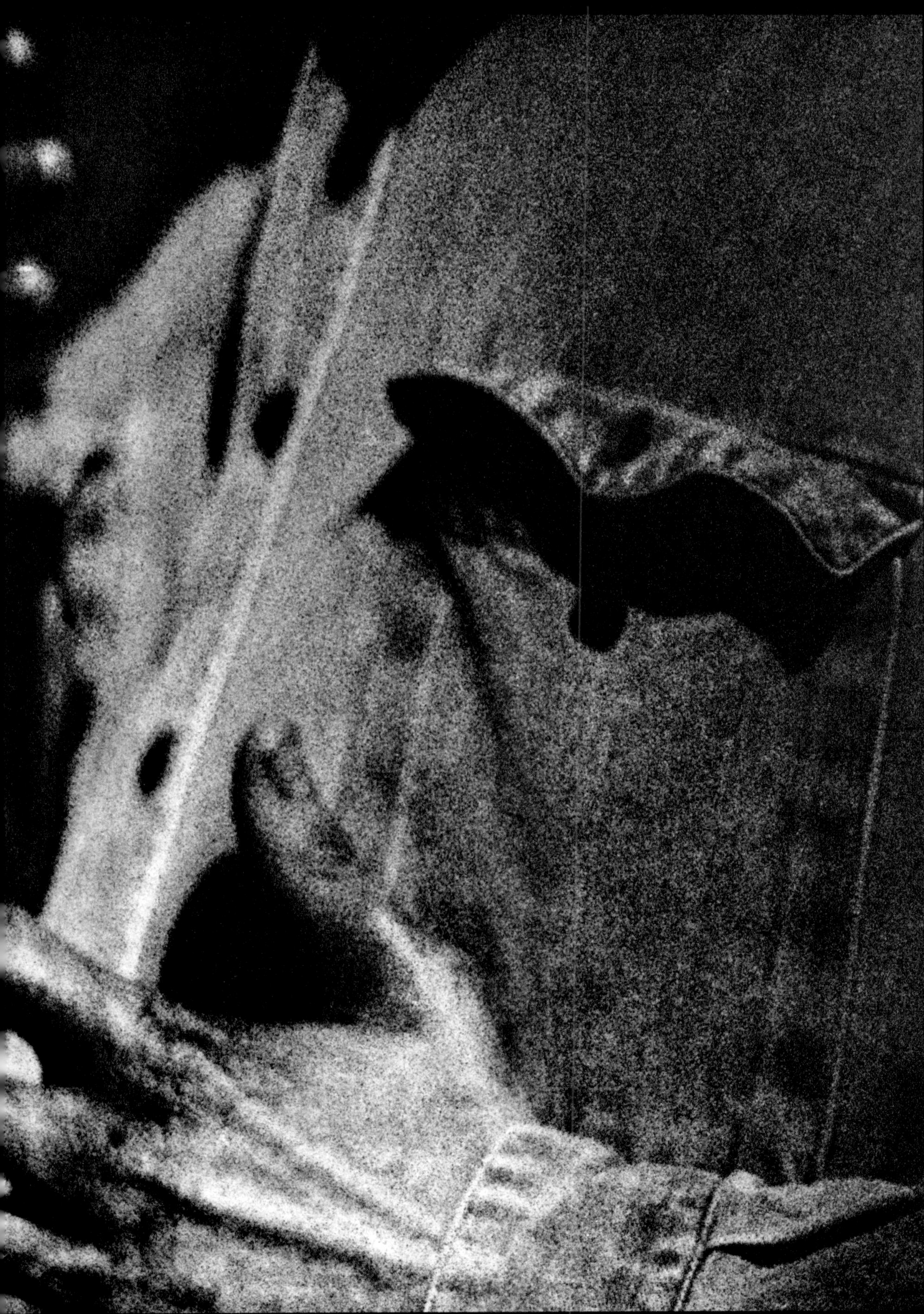

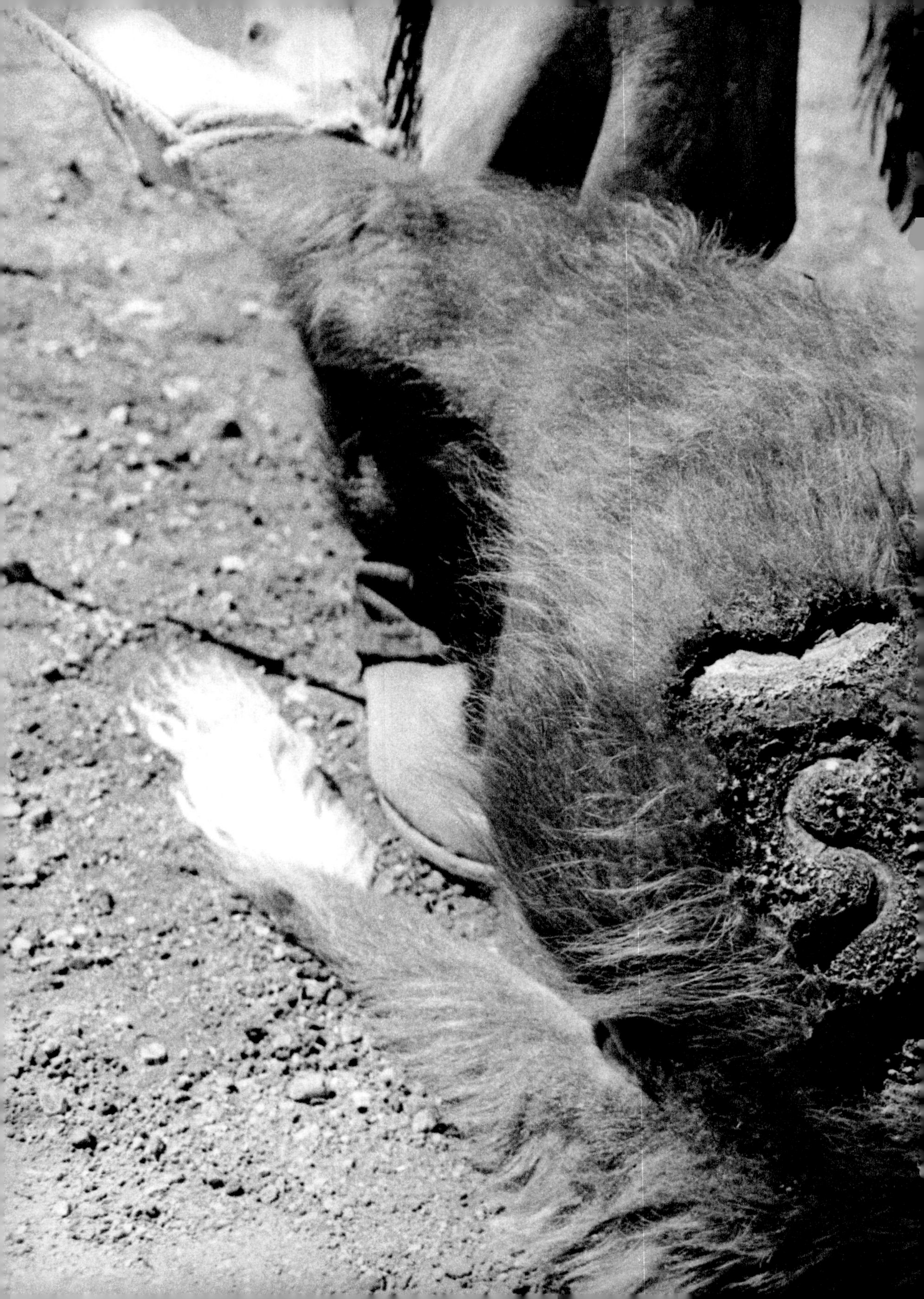

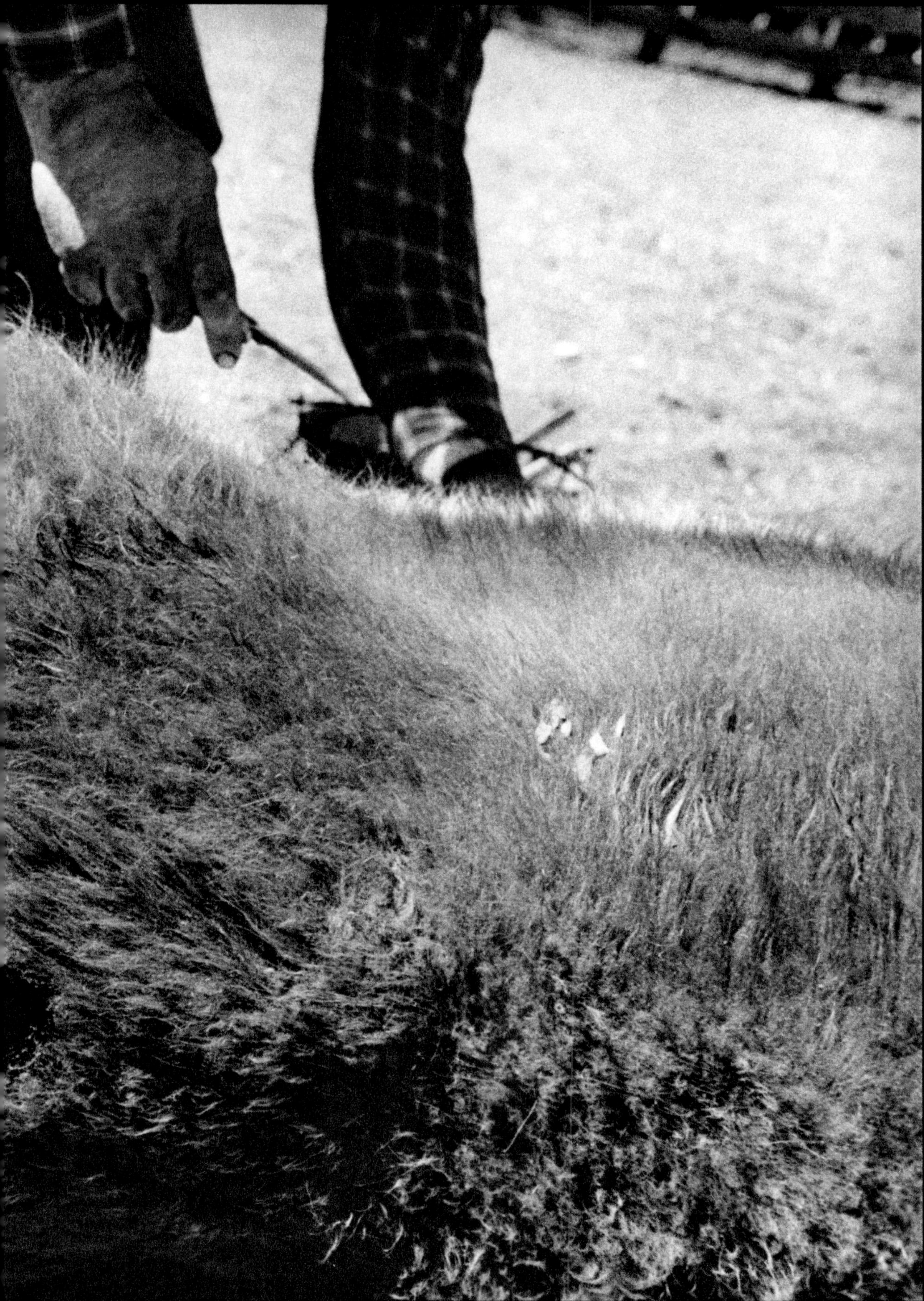

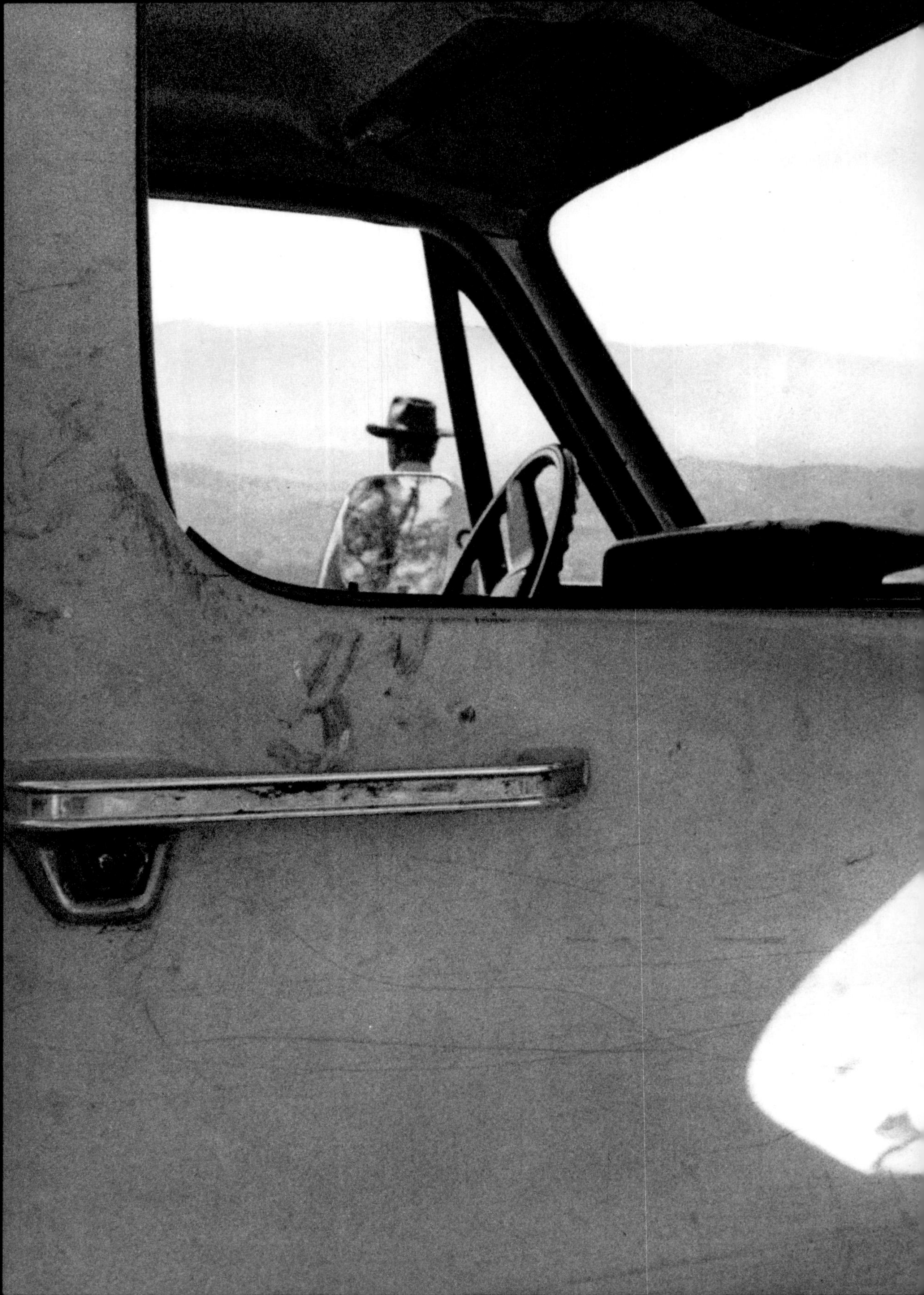

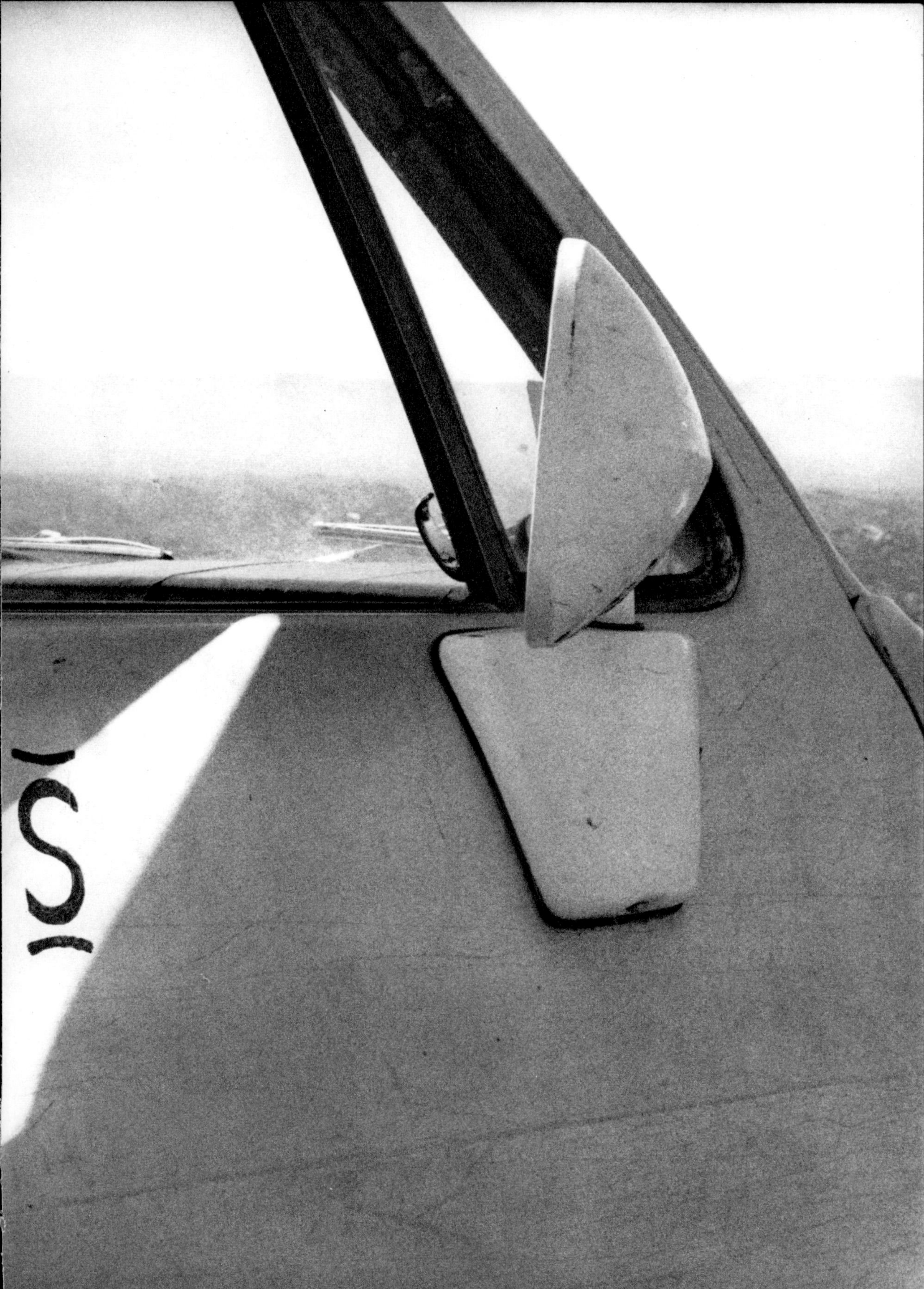

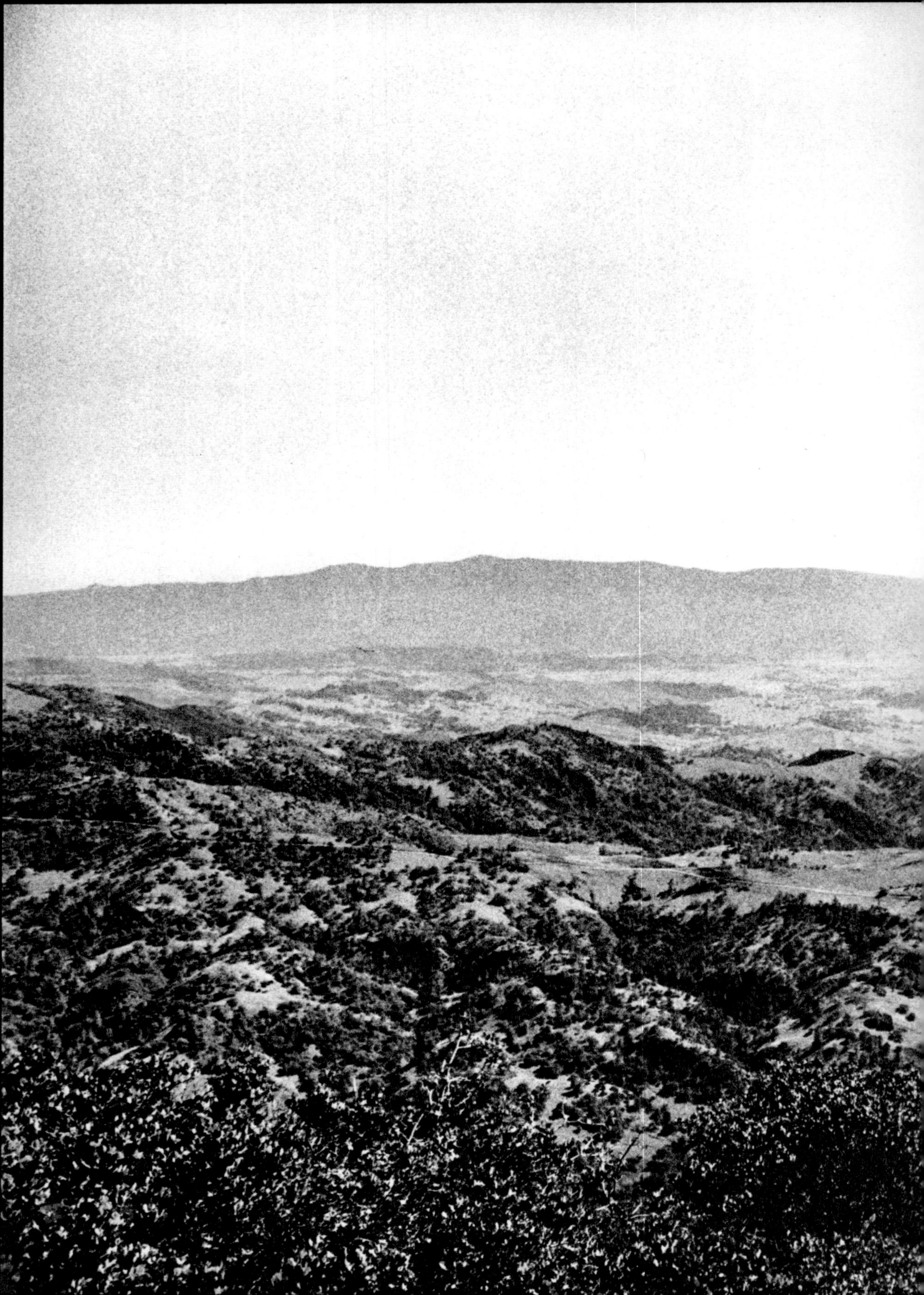

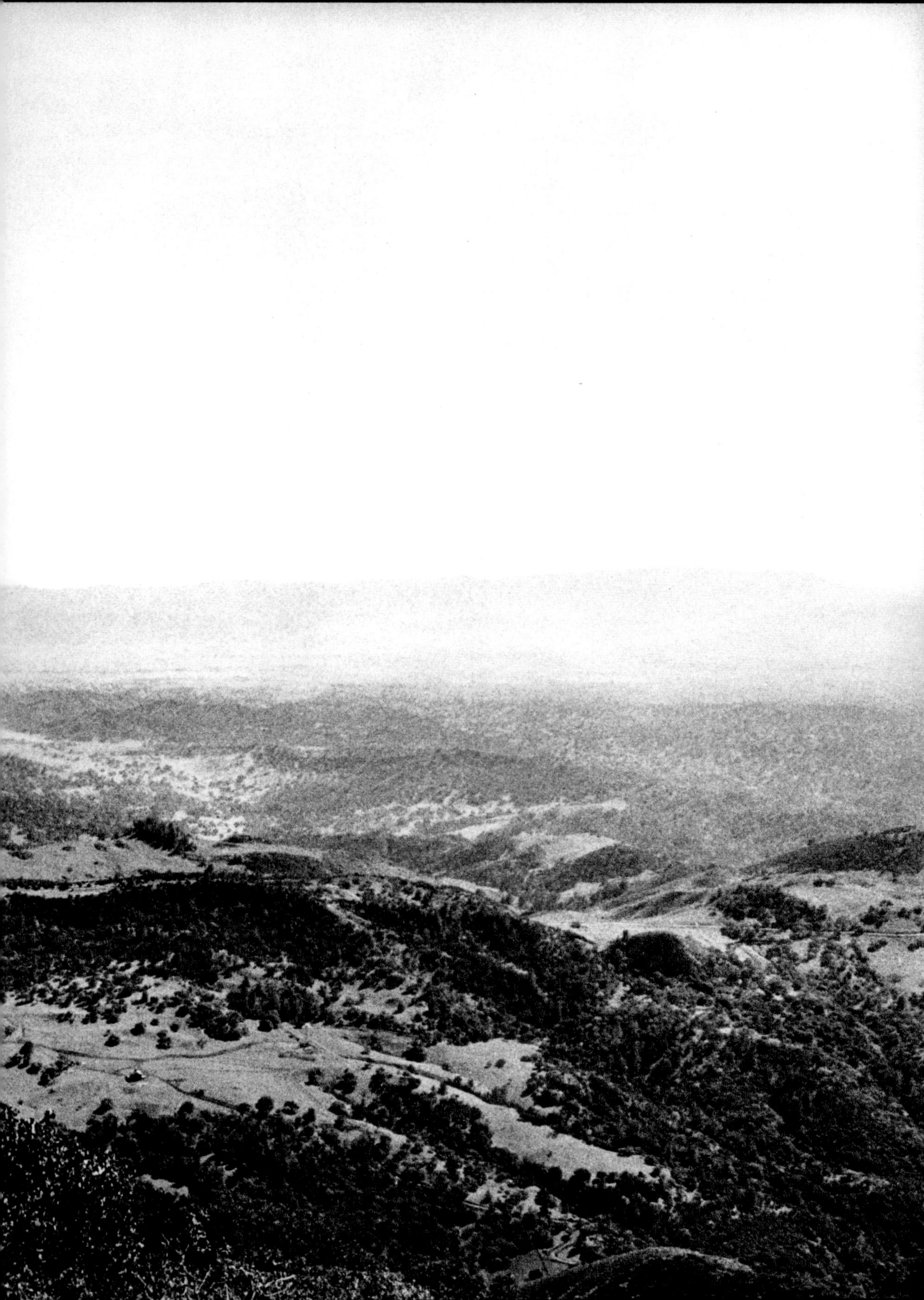

As I drive along Highway 666, I think of the first road I went West on, a town road that ran past my house in the last subdivision in the county, past the stud farm on the other side of the backyard fence — the real border between suburbia's limbo and the cruel paradise of the rural world — past a few older houses, and to just a dirt track that petered out at the beginning of an undeveloped valley.

At the edges of summer days our shadows would stretch out long before us, sliding over the potholes and ditches and only shrinking back to a child's stature on the sleep slopes of the grassy hills on either side of that road. The grass turned gold and parched in the spring, and in the summer gave off a sweet smell something like the scent of summer heat itself. Dust has a smell too, chalky and clean, and in the more arid places even water has a kind of smell — an inviting tenderness in the air when irrigated crops or suburban lawns are coming near. Livestock compounds all these elements into the smell of barnyards. The snakes we caught gave off an acrid smell of alarm, and when we made the mistake of catching a bluebelly lizard by its tail, the tail would break off and flail around for a few minutes before the life went out of it.

There aren't very many images or words about this country, whose elements are rolling hills, oaks, and the simple seasons of wet and dry — mostly dry. Photographically, this California territory between the tidepools and the mountains is a blank: too gentle and modestly-scaled to fit in with the image of the rugged West, it's neither lush or grand enough for scenic calendars. But it's the majority of this westernmost state, stretching from the Tejon Pass north of Los Angeles all along the Coast Range and western foothills of the Sierra Nevada to somewhere in the far north: a great unsung oval framing the agricultural flats of the Central and San Joaquin Valleys. The southern half of this oval was Spanish land-grant ranchos before California became part of the nation, and the wild oats and other grasses that turn so bright and dry in summer came with Spanish cattle. When Richard Henry

Dana arrived in the 1830's, hides and tallows were the region's only exports, and the missions' booming cattle business had been expanded by secular ranchers. It was here that the saddles, chaps, spurs and lariats of cowboys came into being, in these peculiar institutions which mixed the blood of Christ with the blood of cows. The captured natives and their devout enslavers became the first *vaqueros* — a word that itself mutated into buckaroos — and only this strange Californian beginning has been ignored in the otherwise celebrated history of the cowboy.

Highway 666 takes me to New Mexico, where I stay with my friends Jerry and Meridel, in the house Jerry built on land his parents bought. Jerry tells me about when he and his brothers were young and the Western writer Jack Schafer moved in down the road, in the 1950's. Schafer wrote *Shane* before he had ever been West at all, and later on he wrote a book whose hero was modeled after Jerry's brother Archie West, about a cowboy resisting the disappearance of his way of life; like *Shane*, that book became a Hollywood Western too. Now cowboys seem so completely mythical a part of the Anglo West that it is hard to give credence to their reality, but their tools and techniques remain the best means of handling range cattle. So although the epitaph for this way of life has been written and filmed often, it's a little premature.

Last spring I was on a Nevada ranch where a range war was going on — the Feds were stealing cows from the Native American ranchers again. The ranch foreman came in at the end of a day's work and popped a Western movie into the VCR. In the West, the myth seems to have preceded the reality, so that the men who came here already knew what to embody; John C. Fremont in the 1840's and his successors generated mythology out of their adventures with a peculiar prescient sureness. Dime novels about the cowboy West flourished as early as the 1880's (some outlaws apprehended in Montana then had saddlebags full of such dime romances). Since then the cowboy has provided both an iconography of gritty masculine authenticity and a mask of artifice, a mode of being and appearing appropriated by everyone from Eastern Europeans (where a cult of the West flourished under communism) to children, Hollywood actors, and most recently, clothing designers, who have raised these hallmarks of the real to a new level of insincerity. So cowboys present the peculiar problem of a touchstone of authenticity so fenced in by myths and imitations that it is hard to believe there is anything left inside: the fact that cowboys themselves, like the ranch foreman I saw, study the imitations to learn how to be real doesn't make it any simpler.

"Git along little dogie/You're no more to roam/The dinner table is your next home" goes a ditty someone sang to me once. The beef industry in the West could be mapped the way Dante mapped hell, as a series of concentric rings: a niave romanticism of ranch life in the outermost circle, and in the banal center, a slice of beef in a plastic-wrapped styrofoam tray.

Freedom and adventure and scenery are only byproducts; meat is ranching's real product, and meat production is largely a series of steps that reduce and delimit a calf's boundaries so completely that by the time it grows up and reaches the stockyards — branded, inoculated, dehorned, hormone-stimulated, castrated if it's male — it has already traversed much of the distance between entity and product. This is the heart of ranching, and its lifeline to the rest of America. Changing diets and encroaching development make it a disappearing way of life, but the Dantean map might easily be expanded to include a wall of suburbs like the one I started in, well beyond the outermost ring of the ranch itself. Even in suburbia you can find the West if you look hard, follow the roads that go nowhere or just figure out where your food comes from.

Rebecca Solnit

March 1993

For Dee and Helen McVeigh

ACKNOWLEDGEMENTS

Many, many thanks to Jay Dwight and his family, without whose friendship and hospitality this book would not be. Thanks also to those with me along the way: Robert Taugner, Helen Deutsch, Larry Sultan, Rebecca Solnit, Sven, Mark, and Chick, Linda Connor, Barry O'Connell, Pauline Shaver, and Deborah Light, who showed me early on what dirt is really worth.

COLOPHON

This first edition of RANCH by Michael Light is limited to 3,000 casebound copies. The photographs are copyright Michael Light 1993. The text is copyright Rebecca Solnit 1993. This edition copyright Twin Palms Publishers 1993. The printing is sheet-fed gravure on uncoated paper stock. Printing and binding are by Nissha Printing Company, Kyoto, Japan

for

Twin Palms Publishers
401 Paseo de Peralta
Santa Fe, New Mexico 87501
505 - 988 - 5717
ISBN 0 - 944092 - 25 - X

1993

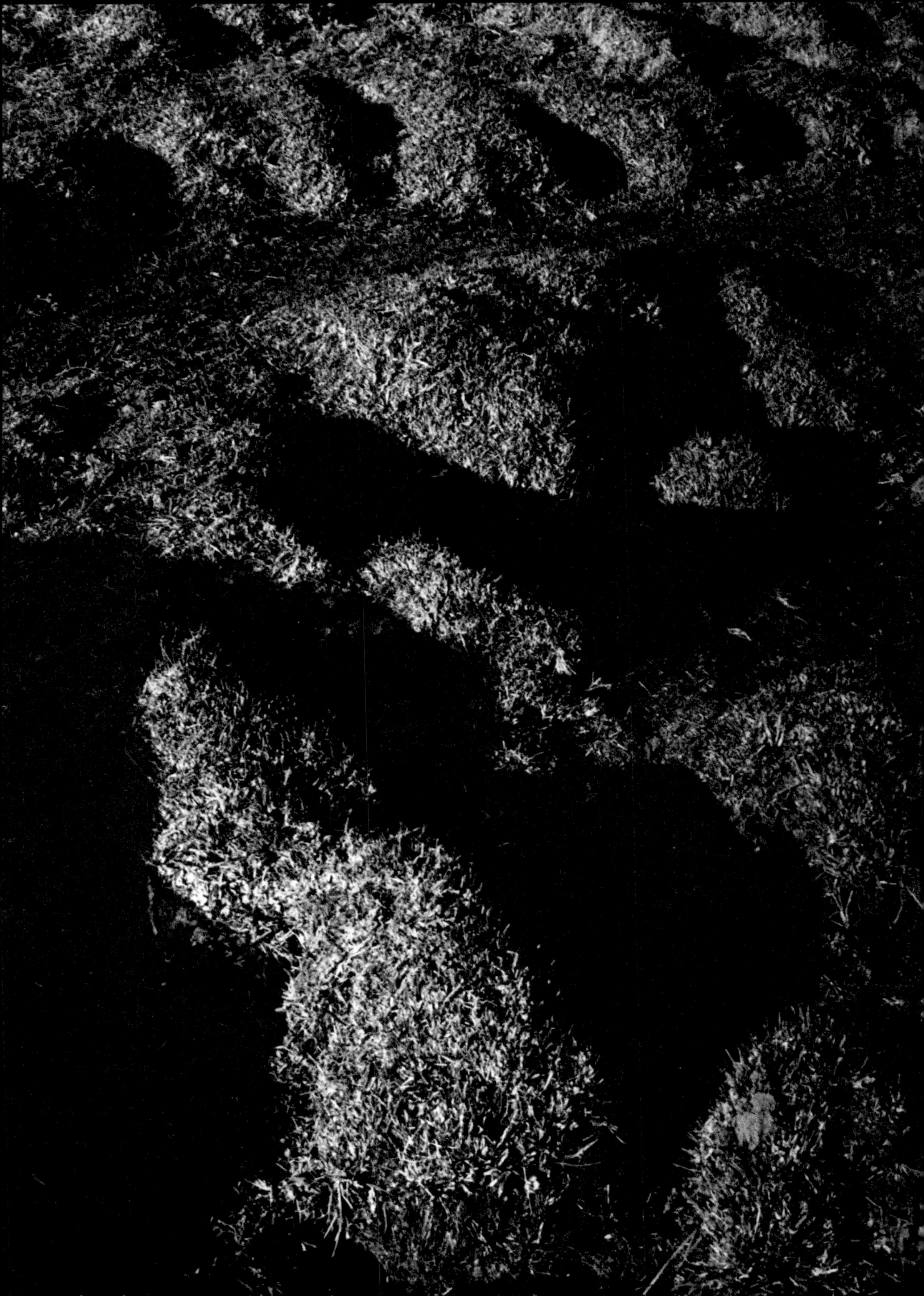